# Energy Choices:
## Opportunities to Make Wise
## Decisions for a Sustainable Future

# QIF Focus Books

# *Energy Choices:*
## *Opportunities to Make Wise Decisions for a Sustainable Future*

Robert Bruninga
with Judy Lumb,
Frank Granshaw,
and Charles Blanchard

*QIF Focus Book 11*
*Quaker Institute for the Future 2018*

Published for Quaker Institute for the Future by *Producciones de la Hamaca*, Caye Caulker, Belize <producciones-hamaca.com> ISBN: 978-976-8142-99-3 (paperback) and ISBN: 978-976-8273-06-2 (e-book)

*Energy Choices: Opportunities to Make Wise Descisions for a Sustainable Future* is the eleventh in the series of *QIF Focus Books* ISBN: 978-976-8142-90-0 (formerly *Quaker Institute for the Future Pamphlets*)

This book was printed on-demand by Lightning Source, Inc (LSI). The on-demand printing system is environmentally friendly because books are printed as needed, instead of in large numbers that might end up in someone's basement or a dump site. In addition, LSI is committed to using materials obtained by sustainable forestry practices. LSI is certified by Sustainable Forestry Initiative (SFI® Certificate Number: PwC-SFICOC-345 SFI-00980). The Sustainable Forestry Initiative is an independent, internationally recognized non-profit organization responsible for the SFI certification standard, the world's largest single forest certification standard. The SFI program is based on the premise that responsible environmental behavior and sound business decisions can co-exist to the benefit of communities, customers and the environment, today and for future generations <sfiprogram.org>.

**QIF Focus Books** aim to provide critical information and understanding born of careful discernment on social, economic, and ecological realities, inspired by the testimonies and values of the Religious Society of Friends (Quakers). We live in a time when social and ecological issues are converging toward catastrophic breakdown. Human adaptation to social, economic and planetary realities must be re-thought and re-designed. **QIF Focus Books** are dedicated to this calling based on a spiritual and ethical commitment to right relationship with Earth's whole commonwealth of life.        <quakerinstitute.org>

*Producciones de la Hamaca* is dedicated to:
—Celebration and documentation of Earth and all her inhabitants,
—Restoration and conservation of Earth's natural resources,
—Creative expression of the sacredness of Earth and Spirit.

# Contents

# Foreword

With *Energy Choices*, Robert (Bob) Bruninga joins a long tradition of Quaker action and witness that speaks to a wide audience about the work of human betterment. Although addressed to members of the Religious Society of Friends (Quakers), this book will be a helpful guide for anyone who is interested in the energy future of their home, business, transportation, or institutional facilities.

We know a tipping point has been reached in energy choices when the Governor of the Bank of England warns investors that the fossil fuel business is a sunset industry with stranded assets, and that smart money is moving to new technologies.[1] The author of this book, along with his colleagues, are well on the other side of this tipping point in researching, evaluating, and mapping out the clean, renewable, and practical energy choices now available to us.

In 2009 Quaker Institute for the Future published a flagship book titled, *Right Relationship: Building a Whole Earth Economy*. While our primary intention was to envision a resilient and sustainable human economy within the context of Earth's ecological integrity, it was pointed out to us that we were also building a case for the "ethic of right relationship" as a fundamental guide for decision making.

The ethic of right relationship has been central to the witness of the Religious Society of Friends since it was founded in 17th century England. An emphasis on right relationship is found in other traditions as well, most notably in Buddhism and in the practices of many Indigenous Peoples who stress living in harmony with the commonwealth of life.

*Energy Choices* is the product of a deep concern for right relationship translated into immediately useful information on clean energy options that are efficient, cost-effective, and well aligned with what Thomas Berry calls "The Great Work" — the integration of human wellbeing with Earth's ecological integrity.

While this book is intensely practical and packed with important information for understanding and making clean energy choices, it is also a guide that addresses an overriding spiritual and ethical issue of our time—the human-earth relationship. Reconfiguring energy use away from fossil fuels to clean, renewable electricity is a big step toward creating a mutually enhancing human-earth relationship. Bob Bruninga shows how we can move in that direction—now!

Keith Helmuth
Secretary, Quaker Institute for the Future

# Preface

Even if we get it from burning millions of years of decayed plant and animal fossil fuels, we are using energy from the Sun. Now we can directly access that bountiful solar energy with solar and wind power systems. Direct use of solar and wind energy, electrification of transportation, and rapid advances in energy storage will revolutionize our future, end the air pollution that is inevitable when we burn fossil fuels, and halt the ongoing degradation of land and water caused by unrestrained extraction of fossil fuels.

Planet Earth is smaller and its resources are far more limited than people often suppose. All the air in Earth's atmosphere, if gathered together, would form a sphere 2000 km across—barely the size of Central Europe. (*Fig. 1, right*).

Figure 1. (*left*) All the water on Earth in one sphere 1400 km across. (*right*) All the air in the atmosphere in one sphere 2,000 km across.[2]

All the water in the world would form a sphere only 1400 km across (*Figure 1, left*). But less than three percent of that total is freshwater. All the rivers and lakes would fit into a very small sphere 56 km (35 miles) in diameter.[3]

Humans have directly modified over half the ice-free land area of Earth.[4] Overfishing has depleted many of the fisheries that provide much or all the protein for millions of people worldwide. Extracting and burning fossil fuels—coal, oil, and natural gas—is increasingly hastening the depletion and degradation of Earth's limited stocks of clean air and water.

Coal and oil combustion is exposing hundreds of millions of people across the world to air pollutants.

Droughts of increasing severity tax overburdened water-supply systems, wildfires destroy homes and forests, and rising seas combined with large storms flood riverside communities and entire coastal areas.

Humanity faces the prospect of meeting the fundamental needs of more people with diminishing supplies of clean air and water, uncertain prospects for agriculture, and more frequent droughts and extreme storms. Triggered by drought, fire, flooding, and famine, deadly conflicts may become more frequent as people struggle with adversity. Anger and resentment are likely to arise as those who presently use the least land, water, food, and energy continue to experience the most adversity.

Those who consume more than their share of planet Earth's resources have an opportunity to promote peace and prosperity worldwide by turning to new ways of harvesting and using energy that do not diminish Earth's capacity to support life. Those who are concerned about the shared future of humanity and Earth are called to transform humanity's relationship with Earth.

Our current lifestyle is simply not sustainable, not in energy, population growth, water use, or the food we produce. Nor are the impacts of high-consumption lifestyles borne equally by all – indeed, those who consume the least are typically exposed to the highest levels of air and water pollution. Each of these concerns is paramount to our long-term survival on this planet Earth, and they can be managed if we have the will to seek truth and balance in our stewardship.

This *QIF Focus Book* deals with the energy part of this life equation and how we can live in a sustainable energy environment now, without sacrificing the quality of our air or our water, or drastically changing our lifestyle. Governments, faith groups and others, organizing and marching, can point the way, but we as individuals are the consumers of energy and the change must come from us.

This book focuses on choices that people can make now. These actions of individuals accumulate and have a huge effect on society, and this cumulative effect is already evident. The electricity use per capita in the U.S. peaked in 2005, for example, and by 2014 it was back to that of 1997, despite increases in gross domestic product GDP.[5] Everyone can participate and every change we make, even changing light bulbs, has an effect.

I went through elementary school in the 1950s in the heady days of science, electronics, and space. Now, at the age of 70 in 2018, I cannot believe that I have lived long enough to see the clean renewable energy solutions here before us. The price of solar power has dropped 100-fold in my adult lifetime since 1970. It dropped 10-fold just in the last decade.[6] Not only do we see the demise of coal at hand, and the amortized cost of clean solar energy now less than half the price of electricity from the utility, but also we are eight years into the electric vehicle (EV) revolution with cars that are better, faster, cleaner, cheaper to operate, and cheaper to maintain than fossil-fueled vehicles when used in local travel and commuting. I can now see a sustainable path to a bright future if we will only consider our choices and make the right decisions with a prepared mind.[7]

However, I have been frustrated by misinformation in the general media, the legacy of our fossil fuel car culture, lethargy of public policy, and ignorance of future possibilities that seem to drive much of popular sentiments in regard to our energy choices. I see good people wanting to change the future by joining groups, marching, writing, being politically and socially active; all in the cause of getting the government and others to change. These are all very necessary exercises, but we must be sure not to let our drive to change others distract us from the individual changes we must each make to bring about the changes so urgently needed.

The government can only give us leadership. But we, with our fossil-fuel burning addictions, are the problem. Instead of wringing my hands over these frustrations, I write this book to present alternatives that individuals have as they make their own energy choices. Although this book focuses on local and simple steps that we can take as individuals to reduce fossil fuel emissions in our daily lives, our individual choices are in aggregate extremely important to all life on our spaceship planet Earth.

Almost two-thirds (63%) of U.S. housing units are single family detached homes that provide many opportunities for both homeowners and renters to reduce fossil fuel consumption and carbon emissions. Residents of multi-unit buildings can reduce their carbon emissions almost as effectively as homeowners by focusing on four out of five major categories of individual fossil fuel consumption. Everyone who drives will have an opportunity to reduce their use of fossil fuels as options for charging electric vehicles are already spreading beyond single-family homes.

This book is the 11th in the QIF Focus Books, a series published by the Quaker Institute for the Future (QIF) with a goal of advancing a global future of inclusion, social justice, and ecological integrity through participatory research and discernment. It comes at a time of transition a decade after publication of the first of that series, *Fueling our Future: A Dialogue about Technology, Ethics, Public Policy, and Remedial Action*.[8] *Fueling our Future* represents the state of climate change concerns in 2008. As the title says, the basic question of that time was which of the available fuels we would use to fuel our future: solar, wind, other renewables, "clean coal" with deep earth storage of captured carbon, nuclear, or bio-fuels. Considering both technical and ethical issues, *Fueling our Future* provided "key information, critical analysis, and thoughtful dialogue on choices for our energy future to assist concerned citizens in their evaluation of public policy and personal choices."

Nearly a decade later, while we still face the emergency of oncoming climate change, there have been some notable changes in the future of energy. Since 2008 the burning of coal has declined over 30 percent and new investment in coal is virtually nonexistent, though part of that is due to the rapid rise in low-cost natural gas extracted by hydraulic fracking.[9]

From 2010 to 2012 there was a 20 percent drop in the use of coal driving the grid which resulted in a cleaner grid, the steepest decline since records began in 1949.[10]

The world-changing event of the 2013 earthquake and tsunami in Japan that resulted in the Fukashima nuclear disaster has had a significant impact on attitudes toward the future of the nuclear industry causing some countries to begin dismantling their nuclear plants.

In 2018 we have very good news of the exponential growth of solar and wind energy, as well as the rise of clean transportation. The resurgence of modern electric cars began in 2009 and in 2018 U.S. automobile manufacturers offer more than 40 electric vehicle (EV) models. Major automobile manufacturers are committing to the future of EVs as the most cost-effective technology that will wean us from the waste and carbon emissions from gasoline and diesel use for routine daily local transportation.[11]

*Chapter One: In Harmony with Nature* summarizes the history of human exploitation of energy sources that leads us to our present situation, which is less problematic than most of us think.

*Chapter Two: Making Life's Energy Decisions* is the crux of this book: the various ways we use energy and the major life energy milestone decisions we usually face anyway. Understanding the clean renewable options available will help us meet those decisions with a prepared mind.

*Chapter Three: Electrification of Transportation* describes the current revolution in electric vehicles that promises to electrify our transportation in the future. Anyone who depends on a car can save money by buying an electric vehicle, whether new or used. For those who cannot charge an electric vehicle at home, charging at work is often a viable solution. Some states have passed legislation encouraging multi-unit property owners to provide charging options and have established pilot programs for on-street charging.

*Chapter Four: Wind and Solar Are Here to Stay* explains the current renewable energy technology. Everyone living in a state with electricity choice can convert easily to wind power purchased from their electric utility. For those living in states that do not offer electricity choice, this chapter provides a rationale for promoting electricity choice. The chapter shows that solar panels are effective even when buildings do not have ideal solar exposure and describes community solar options for those who cannot install solar panels.

*Chapter Five: Energy Decisions in the Home* explains how heating and air-conditioning, hot water, and other functions in the home can be provided in a more suitable way. The heating and air conditioning solutions are not restricted to homeowners—they work for anyone with a window. The solutions for electrifying tools for outside the home are relevant to anyone who works outside.

*Chapter Six: Energy Storage, the Keystone of Renewables* describes the changes in battery technology that are an important part of the current energy revolution. Battery development has implications for future changes in the price structure of energy in the grid. In areas that develop dynamic electricity pricing, such as, time-of-day rates, households will be able to buy electricity when costs are low and sell it when costs are high—an option that can be chosen even by those who are unable to install solar panels.

*Chapter Seven: The Switch to Clean Renewable Energy* shows examples from my own family's 90 percent reduction in fossil fuel consumption through solar energy and EVs while reducing our energy costs by the same 90 percent with no degradation in our lifestyle. Another example is how our Annapolis Monthly Meeting virtually eliminated energy costs and emissions by switching from the propane and grid utility mix to clean renewable solar and wind energy while giving us satisfaction in hosting meetings in a carbon-free and sustainable facility.

*Chapter Eight: What Friends Can Do* summarizes the way forward to clean renewable energy. For those readers interested, the Appendices present technical details of how the technology has developed and why I recommend particular options.

I hope to show that although change to a new clean energy economy might seem daunting at the large scale needed to save our species, it is easy to do as individuals. In our daily lives we routinely face major milestones where we have to make significant energy decisions anyway. As seekers of truth, the prepared mind will know the route to take. It is time to turn our talk into action.

Bob Bruninga
January, 2018

# CHAPTER ONE:
## In Harmony with Nature

*"For this I toil, striving with all the energy which he mightily inspires within me."* – *Colossians 1:29.* Throughout the Bible, this is the only use of the word "energy." This reference is quite fitting as we consider our energy choices.

We only have one planet, spaceship Earth, to carry us through the cosmos. We are now realizing we can no longer consume everything on it for immediate profit and ignore the long-term consequences. We are already beginning to see the consequences of climate change.

As hunter-gathers we simply took what we needed from Mother Earth. Once the local resources were depleted we moved on to greener pastures and began again. When we put down roots, we then expanded the economy into other places where the resources were easier to get, devasting those areas while transporting the resources back home.

What separates us from the other animals on this planet is our propensity to harness external resources of energy to provide for our needs for food, water, and shelter. From tens of thousands of years ago, our source of useful energy has always been to burn something. First we burned fallen wood, then we harvested trees. Where trees were gone, we burned dung and peat and killed whales to burn their blubber. Then we mined coal and finally pumped oil and natural gas. All of these were consumed with vast inefficiency, releasing egregious amounts of carbon dioxide into the air just for the heat we needed for warmth, cooking, transportation, and other processes.

This human expansion, exploitation, consumption, and depletion has worked for tens of thousands of years as humans grew out of Africa and expanded to every habitable corner of the planet. Just in my lifetime the human population has tripled to over seven billion. The ability of our Earth to support this consumption has reached its limits.

In the U.S., our thirst for energy is unparalleled. We individually consume four times more than the world average and, until recently, we have shown no appetite for reducing our lifestyle and consumption. However, we are turning the corner as shown by the decreasing electricity consumption per capita in the U.S. beginning in 2005.[12]

The long predicted peak of oil appears to have occurred in 2016. But it does not matter how much oil and coal is still in the ground, we cannot keep extracting and burning it without significant consequences. This new limit based on how much fossil fuel can be burned per year without risking overwhelming changes in climate is more restrictive than limits imposed by supply.[13]

The new peak oil discussion concentrates on the demand for oil, not the supply. As the demand dwindles and the previously assumed trillions of dollars of economic value of oil vanishes, there will be a significant impact on economic stability over the next decades.

The average home in the U.S. consumes electricity equivalent to about four tons of coal a year to meet its energy demands. The significance of that coal consumption in the eastern states is dwarfed by the unseen 80 tons of trees, habitat, flora and fauna bulldozed into the valleys and streams of West Virginia and other states to get those four tons of coal a year for each of us. This is the equivalent of four railroad cars of habitat lost per year per home in the U.S., an unseen travesty by any measure.

Energy consumption uses natural resources, including land, water, and forests. Three hundred years ago, it might have taken about four acres of forest to yield enough firewood to meet the energy needs of a small European colonist family.

In North America, the colonists did not meet this demand sustainably, but instead cleared and burned millions of acres. Even the earliest photos from the 19th century showed the decimation of forest as Europeans pressed ever westward.

To their credit, humans have found ways to use other forms of energy where practical. We have used the power of falling water for centuries to pump water, to grind grain, and eventually to smelt iron and generate electricity. Even at sea level, the Danes and Dutch developed windmills to pump water out of inundated land to create farms. Seafaring civilizations used wind to navigate the seas. But only in the last century has man discovered the direct conversion of sunlight into electricity and only in the last decade has that process become economical for everyone.

### Sun to Energy

Along with this perilous growth in human consumption, the ability to convert solar power directly to usable energy has also been growing. As excursions into space began, solar cells were new and essential to the space program because it was the only practical way to generate power in space for satellites. The cost then was about $100 per Watt. This was fine for a 0.1 Watt ($10) panel to power my small transistor radio, but for any significant power generation, that meant a $10,000 investment just to light a single 100W light bulb.

As access to space has grown, so has the economy of scale in the development of solar cells. By 2010, the cost of large home solar panels was down to only about $4 per Watt. By 2014, the cost was well under $1 per Watt and the amortized cost of solar electricity was half the retail cost of electricity from the utility. In 2015, the cost of large-scale solar plants for electric utility scale production dropped below the cost of a new coal plant. By 2016 the cost of solar electricity was less than continuing to operate existing coal-fired power plants. As a result, the use of solar systems in 2017 was increasing at nearly 120 percent per year. In 2016, there were 160,000 coal jobs, and that number is decreasing, and at least 260,000 solar jobs and that number is increasing.[14] This rapid growth of solar energy production can help each of us live sustainably.

3

### We Depend on Trees

In addition to fruit and nuts, trees provide shade, absorb carbon, and produce oxygen. With the evolution of photosynthesis, microbes and then vegetation eventually gave rise to the oxygen in the air and our rise as air breathers to populate the Earth. Trees are generating at least half of the oxygen we breathe, the other half coming from plankton in the oceans. Each human on earth needs the benefits of eight mature trees to live.[15] Figure 2 is a sketch of our individual sustainable footprint on Earth, which is a reduction from the four acres needed for firewood energy a few centuries ago down to the size of a house roof for solar energy. We also need to include the other requirements for human life.

A tree area is the average footprint of a mature tree, about 375 square feet. Twenty tree areas can supply the oxygen, and more than two acres of arable land can supply our food. Thirteen tree areas are needed for a septic field, although this can overlap the oxygen and/or garden requirements. Four tree areas are needed for a typical house footprint. Water may come from wells below ground or from rain water on the roof. With the 50 inches of average rainfall a year in the eastern U.S., it would take about 3,000 square feet or eight tree areas of rain water collection and storage to sustain the family (*Fig. 2*).

### Protect Nature Everywhere

The purpose of this homestead illustration is not to suggest we go back to the woods or small family farming, but simply that we should be stewards of all of nature around us, not just the few square feet we might actually own or rent. A third of us live in urban environments but there are still parks and trees, as well as great rural forests and farms to protect. We can have oxygen to breathe as long as we protect eight trees somewhere for our personal air. We can eat if we protect farmland somewhere from being paved over. We can benefit from a few square feet in the community solar garden for our electric production, a few gallons of water from the community water supply, and a few gallons of waste treatment in the community treatment plant, as long as we support the institutions and governments that provide it.

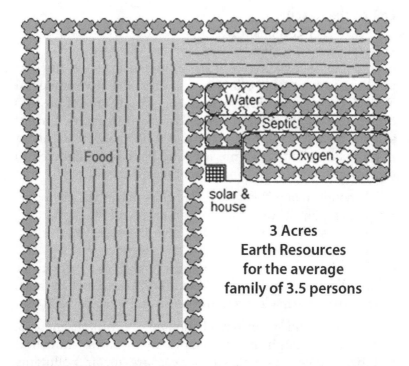

Figure 2. Living on the Land. The sketch above suggests an average family might need about three acres of planet Earth to meet their sustainable resources.

If we assume these three acres per family are needed to sustain life, then we would need 978 million acres for the 326 million people in the U.S. in 2017. But there are only 750 million acres of forest and 280 million acres of prime farmland in the U.S., so it is easy to see that we are getting close to being full. We need to protect these millions of acres of forests, grasslands, and farmlands to sustain our living, our energy and our breathing needs. But we can only do this as a society if we collectively have the will to protect those acres. To achieve this, we must battle the calamity of the human condition, that is, the greed of the few to exploit for personal gain at the expense of all others. Each of us, even if we do not own a foot of land, can work toward the common good and contribute to the protection of our forests, trees, and arable lands. Find a cause, contribute, and preserve your acres somewhere.

### EPA to the Rescue

Fossil-fuel combustion is one of the largest sources of air pollution worldwide. Air quality was getting so bad during the heady days of our car mania in the 1960s that public pressure resulted in the creation of the Environmental Protection Agency (EPA) in 1970. Fortunately, one of the first things they did from 1971 to 1977 was to document the current state of the air in the U.S. Over 15,000 photos were archived in a project called "Documerica" and have been digitized to document the perilous condition of the air then.[16]

Although many think that U.S. cities have bad air today, it is nothing like it was in the 1970s as shown in these startling photographs (*Figs. 3 and 4*) and as seen in images from China today. We forget how far we have come since the EPA and California began to mandate cleaner cars and cleaner energy. Our children have not seen how bad it could be if we roll back the huge progress we have made through clean air and water regulations.[17]

Clean air and water have benefited everyone in the U.S. However, although many people living near industrial facilities today experience less exposure to air pollutants than they did in the past, much work remains to be done. Neighborhoods near freeways, ports, and other places with large volumes of traffic continue to experience higher concentrations of diesel exhaust and toxic air pollutants than do more fortunate neighborhoods. The human costs of air pollution still disproportionately affect those who produce the least pollution. Renewable energy and zero-emission vehicles will especially benefit neighborhoods where progress toward achieving breathable air has been slowest.

Anthropogenic-induced climate change is one of many indicators of human over-exploitation of planetary resources—overfishing, overlogging, over-conversion of natural ecosystems to human-dominated systems. Lifestyle changes are needed in the larger context of the human relationship to the planet. The need to reduce greenhouse gas emissions is immediate, and solutions are immediately available for rapidly addressing this problem. This should be a hopeful message to people who are pessimistic about the future.

Figure 3. International Paper Company mill in1973 (Documerica, 1970s EPA Documentary Series[18]).

Figure 4. 1970s Los Angeles smog  smog depicted in the Honda short film ((Documerica, 1970s EPA Documentary Series[18]).

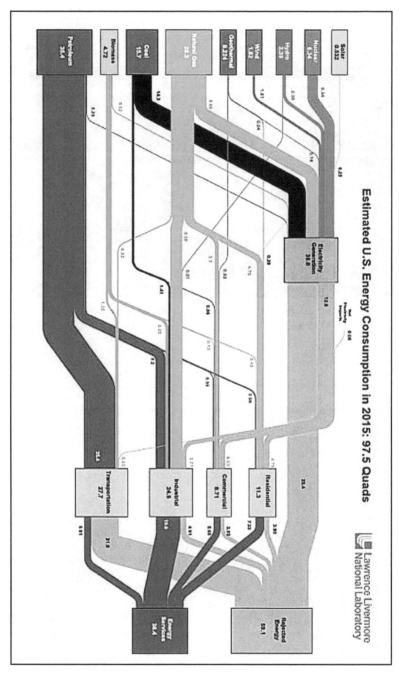

Figure 5. Estimated U.S. Energy Flow Chart for 2016. Adapted from Lawrence Livermore Labs.[19]

## CHAPTER TWO:
## Making Life's Energy Decisions

Almost everything we do demands energy, yet we take the source of that energy for granted and do not usually think of the long-term impacts of our energy decisions. Since the industrial revolution, most of our energy demands have been met by burning fossil fuels. For many people, the changes required to move from fossil fuels to clean energy seem insurmountable or less pressing than our individual day-to-day concerns and daily challenges. However, among the human-managed energy that gives us the lifestyle and products we use, half is under our direct control.

The total U.S energy consumption in 2016 is shown in Figure 5 (*page 8*). Our human energy consumption is about 97 quads per year where each quad is a million-billion British Thermal Units (BTUs). That is equivalent to eight billion gallons of gasoline or 293 billion kilowatt-hours (kWh) of electricity. Another comparison of 97 quads is the equivalent energy of 50 Hiroshima-scale bombs per day. The sheer magnitude of this energy use is beyond our human scale to grasp. The point here is not the huge numbers, but to understand the flow of energy in our current lifestyle.

Figure 5 is to be read from left to right. On the left are the major sources of energy used in the U.S.: solar, nuclear, hydro, wind, geothermal, natural gas, petroleum, and coal. Conversion of some of that energy to electricity is shown in the center. Continuing to the right the figure shows the uses of that energy for residential, commercial, and industrial purposes, shown in the center right boxes.

But not all of that energy provides useful work. In order to use fossil fuels, they must be burned, which produces heat. Unless excess heat can be captured, it is wasted energy, as shown in the bars leading to the Wasted Energy box on the top right. In 2016 two-thirds of the total 97 quads of the energy generated for human consumption is wasted as unused heat (66.4 quads), which comes from burning fossil fuesl or using nuclear energy. The other one-third is useful energy used to heat our homes, operate our cars, turn on our lights, run our dishwashers, produce the food we eat, and make the products we use.

## Our Piece of the Pie

Half of the energy used is in the Commercial and Industrial categories. Although these are beyond our direct personal control, we are still fully responsible for their energy consumption because they provide all the products, processes, and industries to give us the standard of living we demand. But other than changing the way we buy stuff we do not need, there is little we as individuals can do to significantly affect these numbers. However, the Residential category is 100 percent under our control with respect to where we get that energy and how we use it. And the Transportation category is split between our personal use of cars, light trucks, and SUVs (75%) and what our entire economy uses to bring us the commercial and industrial products we crave (25%).

## Transportation

Personal transportation is our largest consumer of energy, our largest generator of emissions and the most inefficient use of all of our energy sources. But, as Chapter Three will show, transportation is easy to fix as individuals when we make our next car an EV or plugin hybrid. We can greatly reduce our carbon emissions and save money too. Nearly half of our transportation consumption is personal cars (*Fig. 6*).

Looking at the energy flow chart (*Fig. 5*), 93 percent of the total energy consumed for transportation currently comes from petroleum; and 79 percent of that energy is wasted as heat in the inefficient internal combustion engine that has driven our transportation sector for the last century.[19] The

10

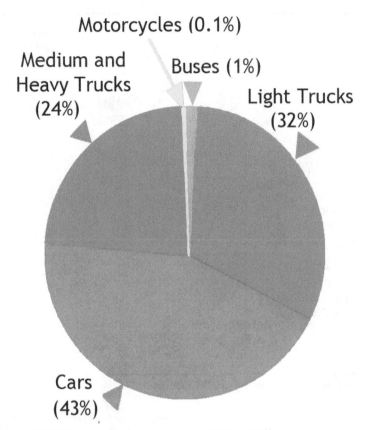

Figure 6. Total transportation energy use (28 Quads)[20]

coming of EVs provides an opportunity for us to reduce this
wasteful use of energy.

### Residential

Once we take out the transportation energy from our
personal energy pie, we are left with our residential use as
shown in Figure 7. Heating the house and heating water
comprise 40 percent of residential energy use. The mix of
residential energy sources has historically been about 60
percent from burning coal and gas, and one-third from
electricity (which in the past was largely generated from fossil
fuels), as shown in Figure 8. But there is no reason to continue
that level of dependence on fossil fuels because as this book

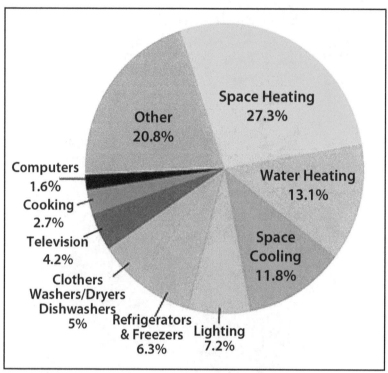

Figure 7. Residential energy use (11 Quads)[21]

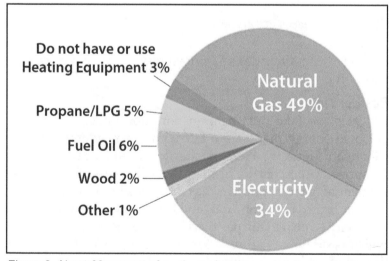

Figure 8. About 60 percent of our home heating and water heating energy comes from fossil fuels.[22]

12

goes to print, 100 percent clean solar and wind energy is cheaper and ready to take over the market.

Clean electric energy can be distributed to our homes as a source of clean renewable energy to power our lives. So, to clean up our personal energy use we need to first switch our fossil fuel energy processes to electricity, while at the same time moving more and more of the sources of electricity away from coal and natural gas toward renewables.

### Growth of Solar and Wind

Wind and solar are the fastest growing energy sectors and coal is declining at a rapid rate (*Fig. 9*). In 2016, renewable resources (wind, solar, biomass, hydro, and geothermal) produced 10.4 percent of all the energy consumed in the U.S. with one-third of that produced by wind and solar. Compared to those data in 2005, there is a 58 percent increase in the renewable share of the energy produced in the United States. Increasing at this rate would bring the renewable energy production to the total energy consumed by 2075.

In addition to this growth, there is another multiplying factor that advantages solar and wind over all the fossil fuel energy sources. As shown in the energy flow chart (*Fig. 5,*

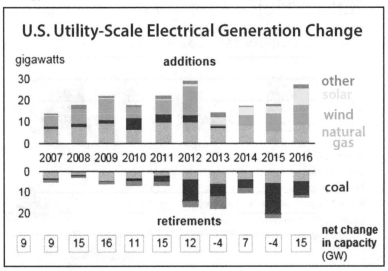

Figure 9. U.S. Utility-Scale Electrical Generation Change 2007-2016.[23]

*p. 8*), two-thirds of all fossil fuel energy is lost as waste heat in the conversion to the electricity energy we actually use, whereas only about five percent of solar and wind energy is lost in the generation and distribution processes. This nearly triples the effective energy produced by solar and wind compared to fossil fuels. And solar energy directly becomes useful residential energy without going through any other conversion processes.

### Life's Energy Milestones

Despite our tendency to avoid facing these big future problems today, it turns out that as individuals we are never more than a few years away from having to make a major personal energy or lifestyle decision. If we approach these unavoidable energy milestone events with a prepared mind, we can be ready to seek the path toward sustainability while also achieving lifelong savings in energy costs and lifelong improvements in our environment.

But if we do not consider our options in advance of these forced decisions, we usually end up making quick, convenient choices, opting for the least expensive immediate solution

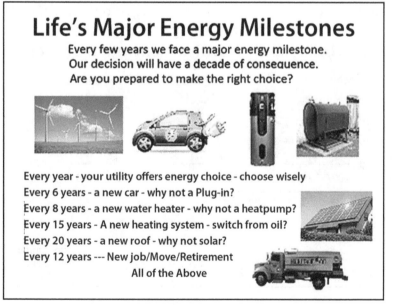

Figure 10. Life's Major Energy Milestones

14

which can cost us two to three times more in the long run, and is almost always worse for the environment. For example, a gasoline car bought today will still be on the road for nearly two decades and an oil or propane heating system repaired today continues burning fossil fuel and spewing noxious emissions. It seems a moral imperative for Friends to at least consider the long-term consequences of such decisions today when already we know so much about how damaging a poor choice can be to our environment and our future. The way we can face these major energy milestones with a prepared mind will be addressed in the remainder of this book.

### Making the Right Choices

The switch to electricity is the first step toward renewable energy.[24] As individuals we can choose to change our electricity source from fossil fuels (coal and natural gas) to clean renewables with a check in the box of a utility form, a subscription to wind electricity from our utility, or the installation of solar panels.

Not only does this switch to electricity provide access to clean renewable energy, in almost all cases, it is also less costly in the long run. If we are lucky enough to have a sunny roof or a tree we can sacrifice to make the roof more sunny, we also gain the self-sufficiency and security of owning our own energy supply for the rest of our lives and being independent of market and political fluctuations in oil, gas, and coal which tend to go in boom and bust cycles while ever creeping upward.

The purchase of solar is the biggest step away from fossil fuels toward clean and secure energy that a homeowner can make. Do you intend to remain in the house for the next 10 to 20 years? Investing in solar is an investment in the long term. It provides a greater return on the investment the longer it is owned. But even if you may be moving, solar is now considered such a good investment that it adds to the value of the house. If you need to sell it in the future, you should easily recoup that added investment. For more detail, see Chapter Four.

While adding solar to your home is the biggest and most consequential step that an individual can make, not all of us are in a position to make that decision right now. The next biggest change we can make that applies to almost every one of us is the electrification of transportation, coming next in Chapter Three.

## CHAPTER THREE:
## *Electrification of Transportation*

Buying a car is our most frequent major energy decision and the one that makes the greatest immediate impact on our emissions footprint. Since 95 percent of U.S. households own a car, nearly everyone is affected by vehicle purchases. On average, individuals in the U.S. make a car purchase or trade about every six years. It is important to consider not just the purchase price, but also the long-term overall lifetime costs and consequences. The average car in the U.S. remains on our roads for 18 years through multiple owners before it is finally scrapped, so fossil-fueled cars purchased in 2017 would still be on the road through 2035, long past the predicted turning point for climate change. Friends with a concern for the environment can simply no longer afford to approach a car purchase without having at least considered the new alternatives and the impact of that choice for our future.

Electric vehicles (EVs) open a whole new paradigm for changing energy consumption to clean renewable possibilities. As this book goes to print in 2018, there is no doubt that local travel, commuting and daily transportation can be done cheaper, cleaner, better, and more conveniently with EVs than continuing our century of inefficient gasoline-burning internal combustion engines and dependence on gas stations and foreign oil. In 2018, there are more than 40 full-sized electric or plug-in-hybrid models available in the U.S. market and at least four manufacturers have announced that all or most of their product line will be electric or plug-ins by 2025.[25]

## Gasoline Hybrids

Hybrid cars were introduced to the U.S. by Toyota with the Prius in 2001 and have been generally accepted as a good step toward improving gas mileage to meet the goals being set by governments around the world. Even though these hybrids still get their energy from fossil fuels, the hybrid electric motor and battery fills in when more power is needed so the engines can be smaller and lighter. The electricity needed to replenish the battery is from regenerative braking and from the gasoline engine.

The biggest waste of energy that could be controlled in any gasoline car is the brakes. Every time you apply the brakes you are throwing away most of the energy you consumed in getting up to that speed. The reason the hybrid gets good gas mileage in town, and stop and go traffic, is that it does not use friction brakes that waste the car's moving momentum energy as heat. Instead when you press the brake pedal in a hybrid or an EV, it switches the electric motor to an electric generator that recovers most of that energy, putting it in the battery for the next time you accelerate. The drag of the generator acts as the brake to slow the car down usually to as low as five miles per hour before the friction brakes kick in. The regenerative braking system that recovers the braking energy and turns the gas engine off whenever your foot is not on the accelerator are the real game changers that allowed hybrids to achieve 50 miles per gallon (mpg) gas mileage, which reduces emissions in half compared to older 25-mpg gasoline cars. Although hybrids have both an electric motor and a battery, they are still fossil-fuel-powered and continue to create carbon emissions, just less.

## Engine Warmup Inefficiences

Just looking at high mpg numbers, however, does not tell the whole story. Until the gasoline or Diesel engine gets warmed up, it is operating at less efficiency. Because the hybrid typically displays the instantaneous miles per gallon (mpg), one can clearly see this. During the first five minutes of my Prius commute, the display shows around 25 mpg; the next five minutes it approaches 40 mpg; and after 15 minutes it is nearing 50 mpg. In addition to this lower warmup efficiency, the noxious emissions are as much as seven times greater during the warm up period because the catalytic converter is not yet up to the temperature where it can convert nitrogen oxides and carbon monoxide to less harmful gases. The car only meets its air quality standards after this warm up period. That is why you are supposed to make sure the car is warmed up before you take it to your state's emissions test. One way to mitigate this problem if you drive for multiple short errands is to always do the farthest one first. This gets the engine up to temperature and then the other short trips are all done with a warmer and cleaner burning engine at higher gas mileage.

The higher emissions as gasoline cars warm up are why EVs are much better than gasoline cars for local driving. EVs have no warm up period and no emissions. They are operating at maximum efficiency as soon as you turn them on. EVs use energy up to three times more efficiently than gasoline cars and all of the braking energy is going back into the battery to be saved for your next acceleration. For short trips under five minutes, using an EV instead of a single gas car is equivalent from an emissions standpoint to taking seven gasoline cars off the road. The considerations of initial high pollution and initial poor gas mileage of gasoline cars are often overlooked in gasoline/EV comparisons.

## Plugins

To simplify the distinction between Hybrid Electric Vehicles (HEVs), which run exclusively on fossil fuel energy, and all the various electric vehicle types, such as Battery Electric Vehicles (BEVs), Plug-in Hybrid Vehicles (PHEVs) and Range-extended Electric Vehicles (REEVs), the only thing that matters for driving on clean renewable energy is whether the car has a plug. If it has a plug, it can run on external clean sources of electricity such as from hydro, solar or wind, or any other form of 100-percent-clean renewable electricity and is called a "plugin." If it does not have a plug, no matter what technology it claims, it is still a fossil-fuel burner.

## Electric Vehicles

Electric Vehicles (EVs) were invented in about 1834 and by the 1890s were outselling gasoline cars ten-to-one. Henry Ford even bought one for his wife because it always worked and she just flipped a switch instead of using the heavy crank to start his model T. It was the difficulty of cranking the early gasoline engines that held back their popularity in those early years. But that all changed in 1912 with the electric starter. It was the end of EVs for a century.[26]

A hundred years later batteries have improved 20-fold in size, capacity, and weight. Electric cars are making a striking comeback that is revolutionizing our transportation sector. This recent re-introduction of the modern electric car began in 2009 with the introduction of the Electric Tesla Roadster. While very expensive, it provided a path for the early adopters to abandon the addiction to fossil fuels and fostered the turn toward clean renewable electric transportation. It could do 0-to-60 in 3.7 seconds, out-performing most gas performance cars, and it could go 245 miles before it needed charging. This was followed in 2010 by the Chevy Volt and Nissan Leaf, both of which were affordable by most in the U.S., and the rest, as they say, is history.

An EV is not a replacement for every gas or Diesel car, but is an excellent replacement for the 95 percent of our transportation that is local and daily. The average daily mileage of drivers in the U.S. is about 40 miles a day (15,000 miles a year). And 98 percent of all daily trips begin and end at home.[27] This kind of travel is the ideal application for EVs because they leave the home in the morning every day with a fully charged battery range, and return home every night for a very convenient overnight charge. These 40 electric miles a day cost about $2 or less, while giving us the convenience of never having to interrupt any daily trip to stop at a gas station.

### Recharging an Electric Vehicle

The vast majority of EV charging is at home using the standard household alternating current 120-volt system. No special equipment is required. Since the car is usually parked overnight, there are at least 8 to 10 hours available for charging. In that time, a standard 120-volt charge cord is adequate to charge the batteries for 50 miles. The range could be extended to 90 miles with the same 120-volt cord plugged into an ordinary electrical outlet while also parked at work. This is referred to as "Level 1" charging which is used for most EV charging while parked at home or work.

Level 2 charging uses 240 volts at 20 to 40 amps to provide three to five times the charging speed as Level 1 charging at 120 volts. At these higher speeds, one might extend the range by plugging in while shopping or visiting other locations. This is the focus of most public charging.

Level 3 is a fast-charging system intended to be located along interstates for long trips. The fastest Level 3 charging takes 20 minutes or more. The usage of the three levels is shown in Figure 11.

Though an EV is not designed to be a distance traveling car, it can charge anywhere there is a standard electric outlet.

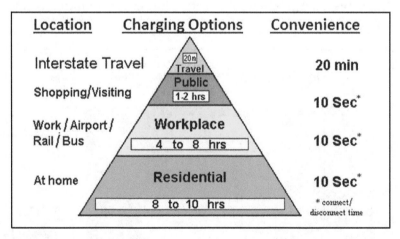

Figure 11. The charging triangle presents the charging options for electric vehicles, showing that the Level 1 overnight charging at home is the most used. The other use of Level 1 charging is at work. Level 2 charging at a public charging station is less often used while shopping or running errands. Level 3 charging is only used on long-distance trips.[27]

21

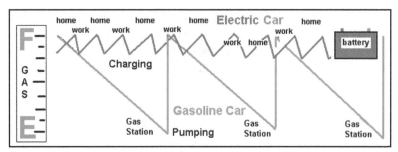

Figure 12. A battery is not a tank.[27]

## *Charging vs Filling a Gas Tank*

Level 3 charging is the only charging process that comes close to mimicking the five-minute gas-tank/gas-station process of the gasoline car that has dominated transportation for 100 years, so people assume it is the only way to recharge an EV. Consumer education is needed to correct this misconception and move forward on the electrification of transportation. EVs do a great job for local and daily travel. Charging happens when parked and not in use. It takes no more than about ten seconds to plug and unplug. As Figure 12 shows, charging an EV is very different from filling a gas tank.

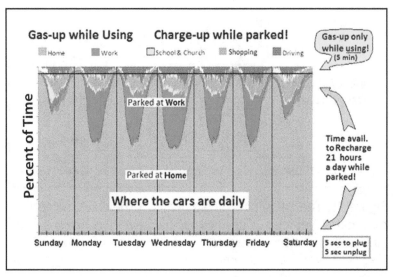

Figure 13. Where the average car is during the week.[28]

22

Notice in Figure 13 that the average car is only in use an hour or so a day for most people. The rest of the time it is parked. For a gasoline car, the only time it can be refueled is somewhere at a public gas station at the most inconvenient time—while you are using it, and you have nothing to do while it is being refueled but stand there and wait. We are so familiar with this method of refueling that we don't realize it is an inconvenience when you compare it to the out-of-sight- and out-of mind EV charging convenience of simply taking five seconds a day to plug in, let the car charge while parked, and five seconds to unplug a full battery every morning!

### Range/Cost Tradeoff

By 2012 most EVs had daily ranges of about 100 miles, which is two-and-a-half times the 40-mile average daily travel in the U.S. By 2017, several manufacturers were producing EVs with over 200-mile range. With almost five times the average miles driven daily, these EVs can fulfill all local and some distance travel needs.

An excellent example of this tradeoff is the expensive and most popular Tesla Model S luxury sedan at over $120,000. While this car has a range of over 250 miles, in 2013 Tesla also sold the exact same car with a smaller battery option for $50,000 less. Choosing this lower cost model gave the same luxury prestige, performance, acceleration, and all the other advantages of driving a Tesla electric while also giving more than three times the average daily range, all at a huge savings of $50,000. Yet, this lower range model was eventually dropped from the lineup because those with money to spend on a luxury car were not interested in the savings, only in getting the best car with the longest range available, whether they needed it or not.

### Cost of Electric Vehicles

Although the EVs re-introduced in 2010 were more expensive than gasoline cars, by 2015 the average cost of all EVs on the market ($29,000 including incentives) was significantly less than the average cost of gasoline cars ($33,000). If we take the luxury Tesla out of the total, then the average cost of electric cars drops to $26,000 which is $7,000 below the average cost of gasoline and diesel cars. In 2016, the average cost of a used EV was under $10,000. Not only do EVs now cost less than fossil-fuel-powered cars to buy and operate, they are also much cheaper to own. They have one-third the energy cost and, without oil changes, mufflers, exhaust systems, radiators, friction brakes, automatic transmissions, catalytic converters, and emissions control systems, they are far simpler and less expensive to maintain. Some estimates suggest that the maintenance cost of an electric car will be only 10 percent of the maintenance costs for a gasoline car.

### Miles per Gallon Equivalent

The concept of "miles per gallon equivalent" or MPGe was developed to make it easy to compare the efficiency of EVs to fossil fuel cars. To get MPGe, the energy content of electricity is converted to the same energy units of gasoline so they can be compared "apples to apples". Operating these modern plugin electric vehicles is equivalent to getting 100 to 120 miles per gallon on gas. This is more than double even the best hybrid models (Prius). But the real value to our future is the elimination of emissions when charged from clean energy and the security and convenience of refueling at home.

### Clean Energy Buyers

One might be concerned that because the electricity grid in some parts of the country may still be powered by nearly 50 percent coal, EVs charged from the grid still have dirty carbon emissions. But most people who buy EVs care about the environment and buy clean electricity from their utility or put up solar panels. A survey by Ford in 2015 showed that 83 percent of their EV purchasers used their own solar system, bought solar or wind electricity from the local utility, or would do so as soon as the utility offered it.[29] Even with a grid running on 50 percent coal, with 50 percent of the EV owners using only clean energy, and the EV using only 33 percent of the energy consumed in a gasoline car, these fractions combine to indicate that the average EV charged in that area will only generate about eight percent of the emissions compared to a gasoline car.

### Battery Swap Technology Myth

As the coming of electric cars was anticipated for the last several decades, a common idea to meet the speed of gas station refueling was the concept of rapid battery swapping instead of the time it takes to charge a battery. This idea carried through to the current California Air Resources Board regulations to encourage EV refueling in the same five minutes as required for filling a gas tank. If a manufacturer was able to demonstrate such speed, they would get bonus points worth millions of dollars of tax and other credits in competition with gasoline cars. With this arbitrary incentive, Tesla made their Model S cars able to swap the entire battery pack in the five minutes required to fill a gas tank and demonstrated it with great fanfare.[30] Since that demonstration, we have not seen a single battery swap, nor will we. If one's battery represents nearly half of their $120,000 investment in a Tesla, it is no wonder that the owner is not going to give up their known $50,000 battery for an unknown battery at some roadside swapping station.

The battery swap concept holds onto the legacy of gas-tank/gas-station thinking. A complete paradigm shift is needed to realize that generally an EV used in its best application of local travel and commuting never has to do public refueling, but simply charges at home overnight or while parked at work. Once commuters get past the century-old legacy of public refueling gasoline cars, the EV is a clear winner with only ten seconds a day to plug and unplug at home.

### General Adoption of EVs

The general adoption of EVs is the next step in the electrification of transportation. Once the general population fully understands and experiences the value that EVs promise for local daily travel and refueling at home, predictions are that the adoption of EVs will be as dramatic as the replacement of the horse by the automobile at the turn of the last century as demonstrated in Figures 14 and 15. It is a challenge to find the single automobile in 1900 in a street full of horse drawn carriages. The car is there in the center of the picture. Yet, just thirteen years later, the 1913 picture shows nearly 100 percent adoption of automobiles and all but one horse drawn carriage (on the left) have vanished.

The average car in the U.S. remains on our roads for 18 years on average through multiple owners before being scrapped. These cars will still be burning gasoline and emitting their toxic brew through 2035. We cannot in good conscience perpetuate these dinosaur burners. Good luck trying to sell one in a few years when they are obsolete. Some countries have already set deadlines after which fossil-fueled cars will no longer be allowed on their streets.[31]

## Hydrogen Fuel Cell Electric Vehicles (HFCEVs)

Another outdated concept is the promise of hydrogen-powered cars. Before EVs became practical, the bet was on the future of the hydrogen economy since the byproduct of burning hydrogen is water — what could be cleaner? But the big problem is the source of hydrogen. There is no source of natural hydrogen on Earth. It is always combined with something else as in water or natural gas, so some other source of energy must be expended to extract the hydrogen.

Natural gas is currently the cheapest source of hydrogen, but obtaining natural gas involves hydraulic fracking with its huge contamination and other problems. It is neither economical nor clean. Producing hydrogen from natural gas generates carbon monoxide and carbon dioxide, creating more carbon emissions. So this source of hydrogen has no place in our future clean energy economy.

Another method of obtaining hydrogen is water hydrolysis, but that process takes much energy from some other energy source to separate the hydrogen from the oxygen. Then more energy is required to compress the hydrogen for storage and delivery. Distributing hydrogen to car tanks is also problematic because no distribution system exists and it is dangerous because of a tendency to explode. Even if the distribution problem is solved, hydrogen is then inefficiently converted to electricity in a fuel cell to drive a motor. But now

Figure 14. 1900: Find the **one car** on Fifth Avenue, New York City, on Easter Sunday. *Source: U.S. Bureau of Public Roads. National Archives and Records Administration, Records of the Bureau of Public Roads.*[32]

Figure 15. Thirteen years later find the one horse in this Easter Morning photo, Fifth Avenue, New York City, 1913.[33]

an EV with a modern battery can be used to drive the motor directly, avoiding all the problems of hydrogen.

Hydrogen fuel cell cars made sense to the "tank-dependent and public refill" way of thinking of energy storage for vehicles 10 years ago, before modern batteries made electric vehicles practical. But since the cost and weight of batteries has dropped 20-fold and the cost of 100-percent-clean solar electricity has dropped 100-fold, there is little to recommend the hydrogen fuel cell car with its three major disadvantages: lack of a hydrogen source, lack of a practical distribution system, and the losses that occur at each of the three conversion steps.

However, hydrogen may eventually have a niche market in the long run. In the future when there is excess renewable wind or solar energy production, the excess clean power might be used to generate free hydrogen from water as a means for utility-scale energy storage. The hydrogen would be kept in tanks where it was generated and then burned in turbines to feed that stored energy back to the grid when needed. Another small niche might be in long-haul hydrogen trucking and railroad transportation hubs where distribution can be done more easily in bulk just to the hubs and along fixed routes.

Hydrogen fuel cell cars are still in the news because decades ago, when it seemed that hydrogen fuel cars might make sense one day, the California Air Resources Board made regulations to encourage their development. Hydrogen car makers still get bonus points for low emissions, just like makers of EVs do. The rules still give three to five times more bonus points to any system that can refuel in the same five minutes. These bonus points are what allows car manufacturers to still sell large numbers of gasoline cars in California, which nominally requires manufacturers to sell larger numbers of zero-emission vehicles than they actually do. As a result, the manufacturers only make exactly the minimum requirement of HFCEVs to sell in California and no more. No one foresaw a decade or more ago how rapidly the EV would overtake the HFCEV concept and beat it in every measure of efficiency, practicality, and cost.

## Transition to Electric Vehicles

This book focuses on EVs as an immediate and personal transition to clean energy because there are models of EVs that are better, cleaner, faster, cheaper to buy, cheaper to operate, and cheaper to maintain than a gasoline car. When purchased by an informed and prepared individual, the transition is inevitable. But why is it not happening faster? Here are some possible reasons:

- Human nature abhors change.
- Our fossil fuel transportation has worked well for a century.
- Human nature clings to past beliefs often in spite of clear and present new information.
- It is hard to abandon the century-old legacy of gas-tank/gas-station public refueling thinking.
- It is hard to grasp the convenience of refueling at home and waking to full daily range.
- Companies are not advertising EVs because of higher profit in SUVs and gasoline cars.
- Dealers actively steer purchasers away from EVs toward higher markup gasoline cars.
- The fossil fuel industry is well aware of this challenge and does all it can to sow misinformation.
- Climate change deniers would have to admit error if the EV is our true salvation.
- Gas prices are cheaper in 2016 ($2/gallon) than nearly $4/gal a few years ago.
- Political upheaval in 2016 brings back the "drill-baby-drill" greed to exploit U.S. oil.
- Many people only buy the cheapest because of limited resources, even though it may cost them more over time.

### The Dealership Bottleneck

When most EV purchasers are asked what is holding back the widespread adoption of EVs, the topic of dealerships invariably comes up. There are many ways that this impacts the sales and growth of EV adoption. Many dealers actively steer potential customers away from EVs to gasoline cars where there is a higher profit margin. Most dealers are not fully trained on the true value of an EV and most salesmen do not own, operate or drive one. Most dealers make most of their income from service, not sales. The sale of an EV with a 90 percent reduction in maintenance is a severe threat to the dealership model. Without the need for frequent maintenance, the dependency on the dealer disappears, especially when many updates can be made directly to the car via the internet instead of a service call.

Over the last century as our complete dependency on the automobile for transportation grew, so did the power and influence of car dealerships that sold and maintained them. This influence has led to laws providing dealerships a virtual monopoly on selling cars and has actually made it impossible for new car manufacturers to enter the market. This grip of the dealership model is unrelenting,

As of this writing, Tesla can only sell cars legally to customers in four states and is fighting the dealership monopoly in fourteen others. Three states with strong dealership lobbies have gone so far as to even enact new legislation to expressly prohibit Tesla direct EV sales to customers in their states.

Since dealerships make the vast majority of their operating profit from servicing and maintenance of the complex gasoline engine and pollution abatement systems, they see the 90-percent reduction in estimated maintenance of an EV as a serious threat to their business monopoly.

### Energy Consciousness
In contrast to the negativity associated with automobile dealers, once someone begins to drive an EV, they tend to have an increased awareness of energy use, the minute-by-minute choices we make during a day's travel, our route, speed, how we dress, and the order of errands. Some consider this sensitivity to energy as "range anxiety," but EV owners soon learn that it is not so much anxiety as it is better stewardship of energy in unity with the environment.

## Chapter Four
## *Wind and Solar Are Here to Stay*
### *Wind*

A vital component of our future of clean renewable energy is wind. While personal wind energy systems might be practical for rural households on the plains or mountain ridges, they are not for urban homeowners because of the low average wind speeds in the majority of urban areas. Except for the roofs of the tallest of apartment houses, urban home owners do not have enough wind to make a system practical. Unless your hat blows off your head almost every time you go outdoors all year long, you don't have any useable wind.

However, wind is entirely viable in large-scale utility projects in remote areas where good wind abounds. These projects are well underway and definitely a choice to be selected from your utility to show support for this very viable renewable energy solution. Investment in large-scale wind generation is exploding. By 2015 large-scale wind generation cost less than coal and less than new natural gas plants by 2017; wind was on track to exceed coal generation in Texas by the end of 2018.[34]

But even if you don't have an ideal location for wind, signing up for wind power from your utility is an easy step toward a personal investment in our sustainable future. Your dollars go to the utility where they are paid to the wind producer who provides the electricity to the utility grid that delivers it to your house. This simple step encourages more wind development.

## Solar

Just as we used the flow of water from a higher level to a lower level for energy over the centuries, the conversion of solar energy to electricity is nearly as simple. It is done at a molecular level inside the atoms of silicon, which comes from sand, the oxidized form of silicon. Sunlight essentially adds energy to electrons in the silicon to bump them up from a lower to a higher energy state. The energized electrons flow through wires to electrical circuits to do useful work while returning to their low energy state. There are no moving parts, just the panels in the sun, and an inverter to switch the direct current from the solar panels into 60-hertz alternating current for household appliances. The electrical connection to our home grid is the same as adding another appliance. An electrician just connects the four wires labeled Line1, Line2, Neutral, and Ground through a circuit breaker in the distribution box. Solar power is simple, which is attractive to Friends.

### Solar Panels

Roofs need care or replacing about every 20 years and they may be our most important energy asset. Any roof decision should fully consider the roof's potential for life-long solar energy production. Once a solar array covers a roof, it greatly extends the life of that roof, which is no longer directly exposed to the elements.

A southern exposure is no longer a firm requirement for locating solar panels. Any angle from east to south to west can work. Pointing southeast or southwest only loses about five percent. And arrays that favor the southwest or even west are a benefit to the grid because they provide the bulk of their power in the later afternoon as the grid load is ramping up for the evening peak consumption. Although a true east or west roof might only get 85 percent of the annual energy that the ideal southern-facing roof can capture, an east/west facing roof can use both sides for solar panels and end up generating up to 170 percent of the ideal south facing roof. Since the limit to most homes going solar is having adequate roof space, the east/west homes have this double roof-area advantage.

But your roof is not the only place for solar panels. Both my home and our Monthly Meeting house chose to install our panels on the ground for a variety of reasons. First was the shade on the roof from favorite trees blocking the sun, and second was to avoid the issue of the present condition of the roof and the unknowns of its remaining life. A new roof would have cost more than the solar and, since it seemed to be in good shape even though it was 20 years old, we considered it had plenty more years before it would need to be replaced. Ground mounting the array avoided the shade from trees and also gave us flexibility in location and pointing direction. It also put the array out front where we could be a standing witness to our commitment to clean energy and not hidden in the back on the roof.

### Shade Trees: Pros and Cons

But what if you have shade? Shade trees have a significant value in reducing air conditioning costs and increasing our quality of life, but there is a trade-off because shade can also detract from solar potential. The area of one mature shade tree (about 400 square feet) is about the same area of solar panels needed to power a small house. While it is disconcerting to cut down a shade tree to install solar panels, studies have shown that each kilowatt of solar panels eliminates about the same carbon emissions as does the growth of seven mature trees.[35] Multiply this by the typical seven kilowatts of solar panels for a home, and putting up the solar array has the same carbon emission reduction advantage as planting an acre of trees!

By another measure, removing one shade tree to install a solar array is about the same benefit as activating 50 trees relative to our overall carbon $CO_2$ emissions reduction. However, besides shade, another major benefit of urban trees is to mitigate urban heat island effects. Roof-top solar makes a small contribution to heat island mitigation since it leaves an air gap below the panels so the roof materials absorb less sun heat: the larger the gap, the better the shading effect. For the typical four-inch gap we see on most homes, it is estimated this effect is about a 10 percent reduction of energy required for summer cooling. But roof-top solar does not provide the habitats that trees do in an already fragmented urban ecosystem.

## Grid Storage

The first uses of solar energy systems were in remote circumstances where an electrical grid was not available. In this case, the system is dependent upon batteries to store the solar energy produced during the day for use at night or on cloudy days. But batteries are expensive. The revolution in the economics of solar energy stems from the use of the grid for energy storage instead of costly and high maintenance battery storage. For off-grid isolated solar power systems, nearly $2 of every $3 invested is in the battery storage system and life-long maintenance, and only one third of the investment is actually generating power. When we eliminate the batteries, and store our daily excess in the grid, then $3 of every $3 invested goes to energy production at retail rates. The revolution of grid-tie solar and the elimination of batteries through net-metering has tripled the value of home solar over the last decade in addition to the ten-to-one reduction in solar panel cost in the same time frame. There is more on the future of batteries for home storage in Chapter Six and Appendix C.

## Net Metering

The grid does not actually store the energy. It just shares our excess generation with someone else who needs it at the time. The mechanism for this sharing is the net-meter. When we draw power from the grid, our meter counts the kWh and at the end of the month we pay for the kWh of electricity we used. With a solar net-meter, when we generate more electricity than we use during the day, our meter can also run backwards, subtracting from our kWh. At the end of the month, if we have used more than we generated, we pay the utility. But if we generated more than we used, the meter is less than where it started, resulting in a net credit in kWh that we can use later. Since it is a one-for-one credit in kWh, we are getting the same retail value for the electricity we generated in excess as we have to pay when we consume it, hence the name "net" meter.

The sun, and our lives are on an annual cycle. At mid latitudes, solar panels will usually produce nearly twice as much energy in the long summer days as in the winter, and our heating and air-conditioning loads vary drastically through the year. Therefore, most utilities only square up any net-metering credit on an annual basis. In Maryland we pay monthly bills showing net consumption, and we do not pay when we have a kWh credit (net retail value). Only at the end of the year, if our meter is farther back than where it started a year ago will the utility actually pay for any net annual excess.

Net metering laws vary and in some states the utility pays retail for the end-of-year excess, but others only pay wholesale. This is fair. The idea of net metering is to let homeowners generate their own electricity at retail rates, but it was not intended to let them build large power generating stations on their lots in excess of their needs.

### Value of Home Solar

Some recent studies consider the relative costs of large-scale utility renewable energy generation versus small-scale home generation. The conclusion was that renewable energy is competitive at all sizes. Large utility renewable energy generation tends to be remote from end users and the bigger the generation operation, the more remote it is. That means larger losses in the distribution. Home rooftop solar is used in the immediate area, so there is no distribution loss. Most of the cost in home solar is in the labor to install it, so those monies stay in the community.

### Solar Cell Efficiency

In almost every discussion of solar power, I hear people say they are waiting for newer, more efficient technology. But waiting for higher solar cell efficiency is a fool's errand. The the space age brought fascinating advances in the technology, which has resulted in the phenomenal reduction in price already noted. As demand and production has driven the cost of solar cells down by a factor of one hundred for the homeowner, the space sector's need for more and more power in space at any cost has increased, driving the development of higher solar cell efficiency. As a result the cost for the most

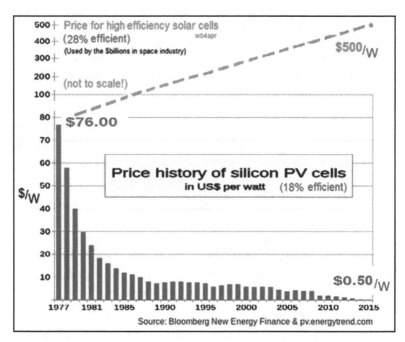

Figure 16. Efficiency of Solar Cells

efficient cells continues to go up. The cost of high efficiency cells is now up to over $500 per Watt while home solar cells are now down below $0.50 a Watt, a thousand-fold difference as shown in the Figure 16.

For this 1,000-fold difference in price, the space industry gains less than a two-fold power advantage, but in space, where satellites cost up to hundreds of millions of dollars each, a few million dollars for the best solar panels is worth it. This huge divergent price difference will never change because the space industry will always pay whatever it takes for higher efficiency on their priceless satellites, while home owners will only pay the absolute minimum. Home solar panels are now cheaper than even a custom glass window of the same size. The biggest cost of solar energy now is not the panels but the labor to install them.

### Peak Power Value versus Distribution Costs
There is a lot of pushback from some utilities, the fossil fuel lobby, and the fossil fuel disinformation campaigns to

convince non-solar customers in some states that, because solar customers are reducing their bills substantially, it must be costing the non-solar customers more. But this is not true. The cost of generating electricity varies minute by minute throughout the day to meet the demand. Although the consumer is only billed the average, say 10 cents per kWh, the actual prices paid by the utilities vary from two cents or less at night to as high as two dollars per kWh during the peak loads on the worst air-conditioning days.

For the residential customer, the nominal cost of electricity is established as the long-term average of these drastic minute-by-minute daily fluctuations. At the high-price times during the day, the home solar net-meter system produces maximum power when it is needed most. But the home owner is still only getting the same 10-cent value when the utility is paying other sources ten to twenty times as much on peak solar summer days. At night the solar customer is still paying 10 cents per kWh used at a time when it only costs the utility a few pennies to produce. The net peak value of the home solar power produced at times of peak grid need more than offsets the modest costs to maintain the distribution hardware of the grid to solar customers. The home solar provides power where it is used in the neighborhood and uses none of the electrical long distance transmission system.

So, it is not the other utility customers that are losing to solar customers. It is the dirtier utility-owned fossil-fuel plants that are losing because the solar generation is contributing during the daily peak when the dirtier utilities could be selling electricity at ten to twenty times the average rate. We need to phase out the dirtiest and most expensive fossil fuel plants because solar systems can minimize the need for them. It is what must happen if we are to clean up the grid and our air.

### Renewable Electricity Supplier

In at least 17 states with progressive energy policies, the utilities offer consumers a choice in energy suppliers. The choice is to continue to purchase electricity from the local grid mix, which can include as much as 50 percent dirty coal, or to subscribe to utility solar or wind energy at a comparable price.

As soon as you open this door to choice, be prepared for an overwhelming barrage of choices, gimmicks, subterfuge, and misleading salesmanship. In Maryland there are at least 40 electricity suppliers vying for our choice. Fortunately, there is a national organization, Interfaith Power and Light, that attempts to sort through all the offerings to find the best price and cleanest power choice for non-profits and people of faith. Contact them as an initial starting point and see what they recommend. At Maryland's 14 cents per kilowatt-hour (kWh) overall electric rate in 2016, the difference to switch our selection from dirty grid mix power to 100 percent renewable wind power was under one penny per kWh, making it cheaper than some of the other offerings. Once the decision is made, switching to clean sources of electricity is as easy to do as a check mark in the box on the utility bill and anyone in a state with progressive energy policy and utility deregulation can do it.

### Community Solar

In areas with abundant trees, only about 20 percent of homes have good solar exposure, leaving most homeowners with little access to the solar boon. Similarly, apartment dwellers, condo owners and renters in multi-dwelling buildings do not have access to the roof or the land on which to invest in their solar future. To meet this need, in some progressive states, the concept of "community solar" is taking root where homeowners without good solar prospects can invest in solar panels in community solar projects where their panels will produce power and that power will be subtracted from the owner's home utility bill. They can even carry the credit to any other home in the same utility service area, or they can sell out if they have to move. This brings solar energy to everyone, not just the 20 percent who own houses with good solar exposure (*see sidebar on p. 39*).

### Financial Implications of Solar

The ninth book in the QIF Focus Book series, *Toward a Right Relationship with Finance: Debt, Interest, Growth, and Security* delves into personal choices in investment decisions and the historical context in which we are now making those choices. It also considers the environmental limits to economic growth,

# Community Solar
## in the Western United States

Electricity produced by sunlight sounds like a great idea, but to apartment dwellers, people with limited financial resources, or home owners living in the shade of tall buildings or trees, the promises of solar can seem elusive. Community solar provides one answer to this problem. Community solar is when an institution such as a church, school, or even a public utility provides the space to place a large solar system on their rooftop or in a neighboring field. Individual households then subscribe to the installation of the system in return for credit for a portion of the power produced by it. If the hosting institution is a nonprofit, they can partner with a for-profit company who will own the system. This strategy reduces the system cost since non-profits are unable to take advantage of available federal and/or state tax credits.

Here in the United States, as this book goes to print, there are 101 such projects that are on-line collectively producing 108.5 megawatts of power. Presently 26 states in the U.S. have at least one project installed and operating. Though the majority of these states have less than three projects, four stand out as leaders. Colorado lists 43 projects and is followed by Massachusetts with 11, Minnesota with 7, and Washington State with 6. California has only three projects on-line, but they are particularly significant because of their size, nearly a megawatt, which makes them among the largest in the world. California is poised to significantly expand the number of community solar projects as part of its continued commitment to the Paris Climate agreement of 2015.[36]

—Frank D. Granshaw

which "leads to considerations of how our society's current realities complicate the challenge of managing household finances in a realistic and ethically responsible way."[37]

In that context the investment in our own home or community solar is very encouraging on several fronts. First, it shifts some focus back on the common good for all of us when we tie our solar system into the grid to help bolster the energy needs of our neighbors during the peak sun and peak loads of the day and, second, it offers us some sense of personal security as well as national security through independence from the global and often devastating grasp of world energy conflict.

It is often said that there is nothing certain in life other than death and taxes, but this misses the third certainty which is utilities. Utilities are now as fundamental to our lives as death and taxes have always been. Now, for the first time, we are in a position to actually solve, here and now, our life-long need for energy with the simple investment in our own solar energy systems.

As observed by Greg Baker, the Energy Minister in the UK, "putting retirement investment into one's own solar system will provide a better financial return than a pension."[38] With ten percent return per year of electricity from the investment for the rest of your life, there is nothing better for future security than investment in a home or community solar system. With an electric car to provide transportation which can run on this clean power from the sun, the retiree is fully independent of the oil industry. The independence and security we get from solar energy and the electrification of transportation gives us freedom from oil and all other fossil fuels.

What could be better for Friends, our community, mutual security and financial wellbeing? Appendix B on "Choosing your Solar" guides the consumer through the process of changing to clean solar energy.

# CHAPTER FIVE
## Energy Usage in the Home
### Heating, Ventilation, and Air-Conditioning

In most climates, heating, ventilation, and air conditioning (HVAC) is the largest and most expensive home energy system. HVAC systems typically have an average life of 15 years. To keep warm we have burned wood, then coal, then oil and, more recently, natural gas and propane. The biggest opportunity for reduction in carbon emissions is to switch from the most expensive of these, oil or propane, to a heat pump. When the price of heating oil rose above $3.50 a gallon in 2015, the cost to heat with oil actually exceeded the high cost of straight electric resistance heating (at the national average of 10 cents/kWh). While the cost of a new heating system is big, that is dwarfed by the lifetime energy costs of fossil fuels.

Natural gas is considered a cleaner fossil fuel because burning it results in half the carbon emissions of coal for the same heat produced, but it is still burning fossil fuel into the atmosphere and hydraulic fracking is used to extract it. Because fracking brings a number of serious problems — pollution of ground water, huge consumption of fresh water, and earthquakes — it has been banned in many areas of the U.S.[39]

The price of natural gas has been volatile because it cannot be easily stored. Generally, it has to be produced and used in real-time, which makes pricing highly dependent on instantaneous supply and demand. Prices in the last decade have been as high as $13 and as low as $2 per million British thermal units. More disturbing are the tremendous leaks of

41

methane involved in mining, extracting and fracking that might be doing even more damage than the carbon dioxide. The greenhouse gas effects of methane are about 80 times higher than carbon dioxide. Some estimates in old cities like Boston indicate that over 10 percent of the gas is lost through leaks. And this does not count the prodigious waste due to the flaming burn-off of leaked methane at oil and gas well heads, or the leaks along the way.

### Heat Pump HVAC Units

The clean renewable HVAC alternative to burning fossil fuels is the heat pump which is two to three times more efficient and has been with us for a half a century. Since heat pumps are electric, they can be powered from any renewable clean energy source, such as your own solar array or utility wind from the grid, which gets ever cleaner over time.

Heat pumps are an excellent opportunity for the switch from fossil fuels to electrification. Electrification allows everything we do to be run from renewable energy since any source of renewable energy can efficiently be converted to electricity for easy distribution. Heat pumps are efficient because they do not consume energy just to create heat but only use electricity to pump heat from a lower temperature source to a higher temperature or vice versa.

We might think that there is little useable heat in outside air when the winter temperature outside is 40°F and 70°F inside. But in physics, heat is measured on the scale of absolute temperature which starts at absolute zero (-460 degrees F). The amount of heat in outside air at 40°F (500°F above absolute zero) is over 90 percent of the heat in 70°F air (530°F above zero). Heat pumps take that lower temperature heat from outside and deliver it indoors at a higher temperature. Typically, a heat pump can deliver usable indoor heat for about half to a third of the cost of providing that heat by burning fossil fuels. Consider replacing your oil or propane heater, or gas furnace, the next time your heating system needs work. (*See Appendix D, Choosing your AC/Heatpump Unit*)

## Fans

Unfortunately, homes built in the last 60 years were designed to be tight boxes maintained by energy intensive heating and air-conditioning. This is in contrast to the whole house fan commonly used through the 1950s. The nighttime low temperature in the summer is 20 to 30 degrees cooler in most areas than the heat of the day. Using a fan to exchange cooler outside air after about 9 pm takes advantage of free cooling instead of burning fossil fuels for the same energy transfer. The art of fan operation involves turning on the exhaust fans in the evening to draw in the outside temperature after it gets below the inside temperature, turning it off in the morning, and closing the doors and windows to retain the cool most of the day.

Though a whole house fan can be added with some expense, it is also possible to do good cooling with one or more box window fans. The key is to set the window fan to exhaust in upper floors or hotter rooms. This removes the hottest air and allows flow control in any other room of the house just by opening windows as needed in any occupied rooms where a fresh incoming breeze is desired. This way, the breeze can be moved from room to room just by opening or closing windows or doors instead of moving fans. As we clean up the daytime summer haze due to all that coal burning for air conditioning, the nighttime skies will get clearer, which leads to lower nighttime temperatures and better cooling with fans.

## New Home

A new home is a major life milestone where energy choices can be considered. Choices such as, what climate, what house, what location, what orientation, what HVAC system are all questions that can be considered for their energy impact. Is the home well insulated, does it have efficient appliances? What about heating and cooling and water heat? What are its solar potentials? Does it have passive solar features like overhanging eves to keep out the summer sun?

Figure 17. Conventional water heater (*left*) compared to two heatpump water heaters.

## *Hot Water*

Heating water is the second largest household energy use in the typical home after HVAC. On average, the typical water heater lasts about nine years, and if your heater is straight electric, it costs about ten times its original purchase price in electric consumption over its life. However, a modern high-efficiency heat-pump water heater may cost three times as much initially, but will operate almost three times more efficiently and save thousands of dollars over its lifetime. Being electric, the heat pump water heater can be powered by clean renewable electricity supplied by the utilities (*Fig. 17*).

The heat pump water heater works by pumping heat out of the surrounding air and using it to raise the temperature of the water. In doing this, it cools and dehumidifies the air where it is located. If the heat-pump water heater is in the basement where humidity is typically a problem, its operation serendipitously acts as a dehumidifier at no added energy cost. If the water heater is in a conditioned living space, this cooling and drying may be beneficial in southern climates where the need for heating is minimal, and cooling and dehumidification is a desired by-product. But having a heat pump water heater indoors near the occupants might be counter-productive in cold climates where the cooling effect would demand more energy from the home heating system.

44

### Lighting

Light-emitting diode (LED) and compact fluorescent light bulbs (CFL) have dramatically saved energy in the last few years. When faced with a burned out light bulb, do not replace it with the same inefficient (under five percent efficient) incandescent bulb that was developed over a century ago. It may only cost $1 but will cost over $60 in energy used over the next few years. Replace it with a $3 LED bulb that will only cost $10 in energy and result in one-fifth the carbon emissions and cost over the same time frame. CFLs contain mercury, so they must be recycled appropriately. They are being phased out and replaced by LEDs

A typical home with about 50 light bulbs can save almost $2500 in energy costs over the life of these new bulbs with this simple energy decision, a 500 percent return on investment. This change to efficient lighting has caused a five percent overall reduction in the entire local grid generation for household energy over the last several years.[40]

# Lawn and Garden Equipment

For many people, manual equipment is cheaper, easier to use, and just as effective as power equipment. For some homeowners and for many employees of yard-service businesses or public works departments, power equipment extends the capabilities of a single worker enough to justify its expense and use. Traditional gasoline-powered equipment, however, is noisy and polluting, and can be replaced by much better alternatives, which are described in this section.

### Lawnmowers

The good old lawnmower and other small tool engines are the worst air polluters of almost all internal combustion engines. They do not add so much more carbon dioxide than cars or other motors, but they add significant toxic air pollution. Information from the EPA according to the People-Powered-Machines web page concludes that the harmful carcinogenic, toxic emissions of a lawnmower used for one hour are about the same as two weeks of daily one-hour commuting in a modern gasoline-powered automobile with a catalytic converter and conforming to all the latest EPA emissions rules (1 mower = 11 cars).[41] Although the net energy usage and contribution to carbon dioxide greenhouse gas emissions from gasoline-powered lawn and garden equipment are small compared to the number of cars, less than one percent of all emissions, their contribution to toxic air is an order of magnitude greater.[42]

Only about a third of the population in the U.S. has lawnmowers and those who do usually live in the suburbs. They are hitting the Earth with a double whammy, first from their energy used in commuting, and second from their large lawns. These tools are used within about three to six feet of the operator, and children, dogs, and cats in the vicinity, all inhaling this toxic brew.

I have long used corded electric mowers and put up with the hassle of the cords because in my very unevenly sloped yard, a very heavy battery electric lawnmower using lead-acid batteries weighing nearly double the weight of a gas mower was an anathema to the concept of lightweight easy mowing. But in 2017 when I needed a new mower immediately, the $350 corded electric push mower was not available at the local tool store. So I paid nearly double the price ($600) for the cordless lithium battery powered model that was self-propelled. Thinking I needed the exercise more than I needed to mow the lawn, I had always shunned the self-propelled types before, but it was the best decision I ever made.

Another life changing advantage of the electric lawn mower is the provision of small headlights on the mower making it possible to easily mow after dark quietly without disturbing anyone.

The new technology of lithium batteries cut the weight to a fifth of those with the old lead acid batteries. The removable battery

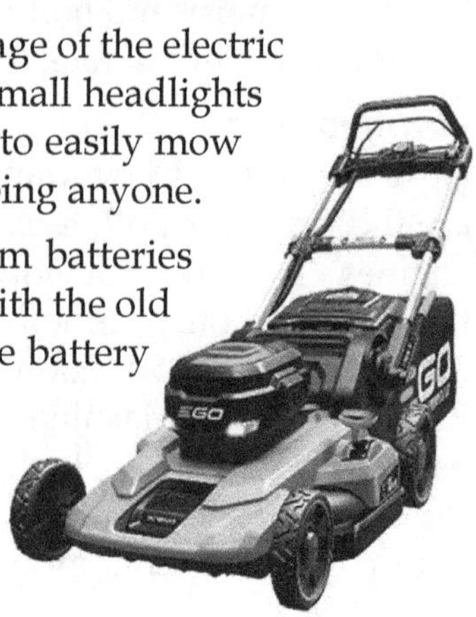

Figure 18. Electric Cordless Lawn Mower with a portable battery weighing only six pounds.

is no more cumbersome than a six-pound bag of flour, so I can leave the mower under a shed and bring the pop-out battery into the house to charge without having to take power to the mower (*Fig. 18*).

## Small Lawn and Garden Tools

Although these small, often overlooked, gasoline engines are only used a few hours a month, their net contribution to air pollution according to an EPA study is 17 percent of all volatile organic compounds, 29 percent of the deadly carbon monoxide, and 12 percent of all nitrogen oxides: the smaller the gasoline engine, the worse the emissions.[43]

Two-cycle leaf blowers, snow blowers, chain saws, trimmers, edgers and cutters accounted for almost half of all emissions of lawn and garden gasoline tools, despite their being used much less than larger four-cycle lawnmowers. All of these tools have excellent electric equivalents and almost all of them are used within one's own yard and easily reachable with an extension cord from your own source of clean renewable solar or grid energy. Or they use modern light weight lithium ion batteries for the ultimate in hand-operated convenience.

## Chain Saw

Of all these home and garden tools, however, the chain saw is a more general tool that is often used beyond the home. It is very useful for cutting limbs and debris at the meeting house and off roads and other areas after storms. Most people assume gasoline power is essential for portability. Yet, with a $99 inverter connected to a car, EV, or Hybrid 12-volt battery, electricity is available that can easily power an electric chain saw anytime, anywhere. When a tree fell across our dead end street, all the neighbors were out there talking of calling the county to come remove it, ignoring the fact that there were thousands of similar trees blocking thousands of similar neighborhoods and it would be several days before they got to us. A few guys had chain saws, but some would not start and others did not have gas or oil. So no cutting was taking place. I stepped over the tree, walked to my house, got my lightweight electric chain saw, came back, plugged it into the inverter in the trunk of my Prius and cut the tree. Compared to the stinky, messy, oily, noisy, toxic, hazardous gasoline chainsaw, I love my electric every time I use it. Since I don't have to carry oil and gas, it is easy to keep in the back of the Prius (with an extension cord) to serve occasional cutting needs at home, at the meeting house and on the road without any smells or leaks.

The other major advantage of electric power tools is the significant reduction in noise. Almost every weekend there is no silence in my suburbia to enjoy the birds and breezes. Some neighbor of the hundreds within earshot is always powering up one of their tools and when they are done, another one begins. It never ends. Not only the large number of neighbors, but also the number of gasoline powered tools. My neighbor prides himself on having a gasoline power tool for everything. Lawn mowers, trimmers, edgers, mulchers, chain saws, leaf blowers, generators and even snow blowers. Multiply by a hundred neighbors and there is no silence in my neighborhood.

### Snow Blowers
The winter tool of noise and toxic emissions is the snow blower. The cacophony of blowers that start up the day after a heavy snow is deafening. Yet my tiny electric blower takes up only 20 percent of the storage space of my old gas one and it runs every time I plug it in, compared to the half day in the cold it usually took to get the gas blower running. Combined with my standard 120v EV charging outlets in my driveway and on a pole down by the street, I can do my whole driveway and cul-di-sac with just a single extension cord. And even if there is no power, a common occurrence after a storm, my home solar or the inverter in my car trunk is always available.

Switching to electric tools, appliances and garden equipment is part of your personal steps toward electrification in preparation for the clean renewable energy economy that is right here before us.

## CHAPTER SIX:

# Energy Storage, the Keystone of Renewables

Cost-effective energy storage is the fourth component of the renewable energy revolution. First, the exponential growth of cheap, clean solar and wind power has been displacing coal. Second, the exponential growth of affordable EVs, both public and private, has begun to displace diesel and gasoline engines. Third, the eightfold drop in the cost of big batteries fueled the exponential growth of EVs and is now supporting both home and utility energy storage. This chapter addresses the coming role of energy storage in the seamless integration of renewable energy into the electricity grid and its carryover into home energy storage systems.

Understanding the role of the instantaneous price of electricity driven by supply and demand is key to understanding the implications of battery storage for solar and wind energy in the grid. On a minute-by minute basis, there can be five, ten, or even twenty-fold fluctuations in instantaneous wholesale electric pricing as the grid managers buy and sell electricity to balance supply and demand throughout the day. Despite this highly variable pricing of the varying supply and demand, the average consumers pay a fixed price per kWh on their monthly bill. This fixed price is negotiated between the various local public service commissions and the utilities to cover the average annual cost of electricity, utility profit, and the cost of transmission, distribution, and billing.

## Dynamic Electricity Pricing

For the last century, electricity was cheapest at night when usage was low and power was provided by baseload coal and nuclear. The typical overnight price was around 2 to 3 cents per kWh. But then during peak air-conditioning days, this price could jump to 20 cents, 50 cents and even over a $1 per kWh as demand exceeded supply at any given instant. Customers in the Maryland market, for example, could get a hint of this variability on their bill as everyone with the new smart-meters was offered a "peak reward" of $1.25 per kWh for any electricity they did not use on certain designated peak summer air-conditioning days. This proved in writing to every consumer how much electric prices vary during the day and how high prices could get during peak demand such as on a hot summer day.

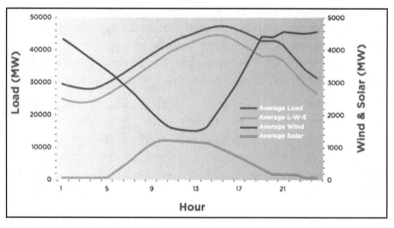

Figure 19. Wind and solar can complement each other to produce a more even output over the 24-hour day in some areas.[44]

## The Complement of Wind and Solar

Fortunately, the variability of renewable energy supply is mitigated in some areas of the country where the wind blows more strongly at night and complements solar power production during the day, as shown in Figure 19. Texas has its own independent grid and vast areas of wind and very high levels of sunshine. Detailed studies in Texas are revealing that there is value in developing places that have wind profiles that produce power at times that are most opposite to solar. It might be better to align some arrays toward the southwest to better match the usual consumer afternoon/evening demand curve, rather facing all solar arrays south to maximize total energy production. These complementary placements reduce the battery size needed for utility-scale storage.[45]

## The Daily Cycle

Solar energy production is growing everywhere but not all states have ample supplies of nighttime wind to complement their solar. In 2017 there was so much solar in California that during a sunny day in the spring or fall when no heating or cooling was required, the supply of electricity could actually exceed the demand. The result is that the wholesale price of electricity goes very low, even much lower than the price at night, which is usually the lowest. Figure 20 shows this for 11 March 2017 in California. Notice how the wholesale price for electricity was a negative one cent or so from 7 am to 3 pm on this particular day, meaning there was so much solar that utilities were paying big customers to use power.

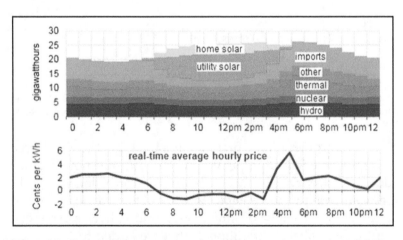

Figure 20. Electrical power sources throughout one day (March 11, 2017) of electrical generation operation in California.[45]

## The Duck's Back Curve

But there is also something else significant on this chart and more visible on the next graph. Solar in California now provides most of the peak daytime energy in the spring and fall when there is the least heating and air-conditioning to load the grid. But it also creates a huge problem for the utility when the sun begins to go down later. The evening has always been the peak load around 6 pm when most businesses are still open and many of the consumers go home to turn on all their AC units and lights, and plug in their EVs. As the sun drops and solar power decreases in combination with increasing load, demand increases at such a rapid rate that the utility is seriously stressed to meet it from about 4 pm to 6 pm.

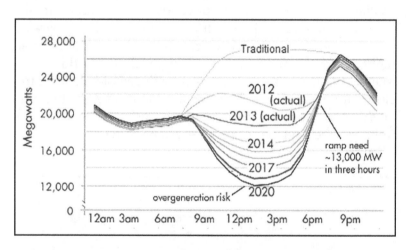

Figure 21. The Duck's Back Curve shows the net load in March each of the indicated years. Adapted from U.S. Energy Information administration.[46]

This problem was first documented by the California Independent System Operator (CAISO) in the now famous duck's back curve (*Fig. 21*). In the years before solar, you could see a smooth rise in demand from about 8 am to an afternoon peak and then a slow decline toward midnight as can be seen on the line labeled "traditional" that we added to Figure 21. By 2012, there was enough solar in the grid to start driving up the supply and driving down the midday demand. By 2013 the solar supply roughly equaled the added demand giving a flat line. But in later years the supply from solar more than equaled the demand in the day. The resulting curve begins to look like a duck's back and hence the name. Meanwhile, the demand from consumers in the late afternoon and early evening remained unchanged. This led to an ever-increasing load on conventional generation that occurs from about 5 to 7 pm. Although this dynamic on the grid is the most drastic in the spring and fall months when there is no heating and cooling on the grid, it does represent the growing significance of grid dnyamics to energy costs.

Meanwhile, the residential consumer sees nothing of this supply and demand costing information. In 2017 the average U.S. customer paid a fixed rate of about 12 cents per kWh because the regulatory bureaucracy, existing Public Service Commissions' rules, and billing infrastructure have not caught up to this new dynamic reality. Since consumers always pay the same flat rate for electricity used any time of the day or night, they have no incentive to consider when they use power to mitigate these fluctuations in price. They are oblivious to when the utility is paying ten times as much to produce electricity.

### Utility-scale Energy Storage
The solution is to have the utilities store energy when it is cheap and use it when it becomes more costly. Historically, the only practical utility-scale storage capability was to pump water uphill into reservoirs and then release it later to

Figure 22. Tesla's largest-to-date utility scale battery in Australia suitable to power 30,000 homes.[47]

generate electricity. But these are massive investments and slow to respond. With the eight-fold improvement in large EV batteries over the last decade, the utilities see the future potential of batteries for this type of storage. An excellent example of utility-scale battery storage that was installed by Tesla in November 2017 near Adelaide in South Australia is shown in Figure 22. The battery was added to the Hornsdale Wind Farm near Jamestown to help alleviate some of the state's severe energy issues.

So now it is battery storage that can save the day. Though the earlier chapters on home solar stressed how un-economical batteries were for self-storage of solar power, the Duck's Back Curve may change all that in many states. It is changing the *status quo* rapidly in California and over the next decade is likely to drive similar change in other states across the country.

This change brings a significant opportunity for investment in peak energy storage. If this dynamic pricing is passed along to the end users, anyone connected to the grid will be able to participate in this energy storage market. This would significantly change the economics of home storage because the home would not have to have a huge battery to store daytime solar to run all night long, but to simply store enough daytime energy to help the grid meet the peak late afternoon two-hour ramp up. It would be worth it to the utilities to pay the battery owners a higher rate for electricity during the two hours of extreme increase in demand because other sources of electricity to meet that surge would cost as much as ten times more.

### Smaller Batteries
Since the only time the electricity value is so high is during the two-hour peak ramp-up, the solar customer does not need total battery capacity to store all the daytime power, but only for the two-hour grid ramp up when it is the most valuable. This reduces the battery investment to only one-eighth of what would be needed for the full 16-hour overnight winter storage. The combination of the five-fold increase in electricity value, the decreased battery size needed, and the lower battery

cost should make home battery storage very practical in the future. If the pricing structure would allow selling power at a profit, then the home backup power problem would be solved as a free by-product!

The five-fold increase in the value of energy stored in a battery and supplied to the grid during peak demand is completely independent of where you get that energy. So the same economics for the battery apply whether the customer is a solar customer or not. Solar customers would charge their home battery during the day when over producing. But non-solar customers could also store grid power in their battery whenever the grid instantaneous price dropped low. In the future the lowest priced power from the grid in solar-dominated states will probably be during the middle of the day due to all the other solar sources. Maryland already recognized this potential and in 2018 began to provide credits and rebates to customers who installed home battery storage.

### Tesla Powerwall

Elon Musk, the Tesla pioneer, pushed Tesla to develop mass market EVs for the new clean energy economy and merge with Solar City to build clean energy solar in homes all across America.

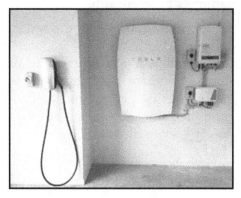

Tesla has invested billions in the Tesla Battery GigaFactory to meet the future

Figure 23. An EV charging cord, the Tesla Power Wall ® battery and the grid interface.

demand of EVs and future homes for batteries. As this book goes to print in early 2018, the 14-kilowatt Tesla Powerwall battery with supporting hardware to integrate with the home costs $6,200. This 14-kWh battery matches nicely the average home solar output of about seven kilowatts for the two hours that electricity prices spike.

To analyze this investment, consider a battery storing 14 kWh a day times 365 days for a total energy supply of over 5,000 kWh per year. At twelve cents per kWh, the battery energy value would be only $600 per year, which would take 11 years to amortize the cost at today's net-metering rates. But by 11 years, the battery would be worn out. But if the utility were paying five times the average rate for those two peak hours every day, the payoff would be $2500 per year and the battery would pay for itself in only three years. As long as the battery would last more than three years, it would be a good investment.

Fortunately, battery and solar costs are likely to continue to go down, and the value of storage to the grid in the evening ramp-up will probably go up. So if the public service commission, the utilities, the politicians, and the bureaucracy catch up to this new reality of electricity supply and demand in the age of solar, and pass the dynamic value along to the end consumer with smart meters, then home storage will definitely be in our future.

The good news is that the battery decision is independent of the solar decision. One can invest now in solar without any consideration of a backup battery system and add the battery storage anytime it becomes economical, when the specific local variables, pricing and politics all align to make that a good investment too. See Appendix C on Choosing A Home Power Backup System.

# Chapter Seven
## The Switch to Clean Renewable Energy

EVs and solar together are a perfect marriage for our future of clean renewable energy. Just covering a single parking spot with solar panels can fully charge an EV everyday on clean renewable energy from the sun to meet the average 40 miles a day usually driven in the U.S. Solar panels also keep the car cool, cutting down the AC energy needed when driven.

Figure 24. Just nine solar panels in Sunny California can fully charge 40 miles a day of transportation. In other states, 12 panels might be needed to do the job.

## My Family Example

My family's interest in renewable energy transportation began in 2007 when Nissan and GM announced their plan for full size standard production EVs by 2010. Since my senior project at Georgia Tech 37 years earlier in 1970 was to build an EV from an old VW, some batteries and an electric motor, these new developments re-kindled an old flame.[48] Between 2007 and 2008, I converted three salvage Priuses from junkyards to a plugin hybrid and a pair of eclectic do-it-yourself publicly visible statements on economical electric transportation. Since the range was short, and I was not allowed to plug in at work to re-charge, I added solar panels to the roof for two of them and put a sign on the side that read, "Solar Plugin." Eight hours in the sun would add about eight miles of free electric miles for the ride home. But in the winter with low sun angles, not so much.

When we bought our house in 1990, I had been interested in energy efficiency and the utility was offering time-of-use (TOU) metering that cut the daytime price of 10 cents per kWh, to only 2 cents at night. With that five-fold difference, it seemed that I could charge batteries cheaply at night from the utility for use in the day when the cost of electricity soared. After doing the calculations, I was shocked to find that the cost of battery replacement every five years would eat up any cost savings. The subtle lesson to me was that even if electricity was almost free (2 cents or less from solar), the cost of batteries and maintenance would eat up any benefit and the upfront cost could never be amortized. This conclusion kept me out of solar for decades.

Still solar kept getting cheaper. My epiphany came on the second Saturday in August 2010, when I realized that this modern revolution in home solar had nothing to do with batteries. Modern solar was grid-tie, meaning excess daily solar generation is stored in the grid by pushing the electric meter backwards and when we need power back at night the meter goes forward. This revelation immediately upended my preconceived notion of home solar based on the high cost of batteries. Although our roof was surrounded by trees, we had a clear area at the back of our lot and by spring of 2013 we were finally on line with an eight-kW array providing all of our annual electric needs and saving over $1,400 a year with nearly zero electric bills and without any batteries.[40]

## Geothermal Heat Pump

With our electricity source thus converted to solar, our biggest remaining carbon footprint was the typical 1,000 gallons a year of heating oil we were using. The price had risen to $3.50 per gallon. Burning oil just for heat seemed unforgivable, so we switched to a geothermal (ground source) heat pump. By the fall of 2014 the ground-source heat pump system was installed and the nearly 80-year-old black, sooty, dirty, stinky, leaky, greasy oil boiler was gone!

Because the heat pump is very efficient it generated almost two-to-three times the heat from the same amount of energy. It eliminated the $3000 annual home heating oil costs and only added about $1500 to our electric bill. We met this by doubling our solar array to a new total of 16KW. After that, our highest electric bill, our total energy costs except gasoline, was $30 in February, while the bill for all the other months was only the $8 minimum utility account fee.

Figure 25 (*next page*) shows the decline of our fossil fuel consumption from almost 3,000 gallons a year (gasoline, coal equivalent electricity and oil) to about 300 gallons a year by 2015 by switching to the Prius vehicles, geothermal heat pump, and solar panels. I found a low cost EV to replace one of the Priuses which further dropped our gasoline use to only a few hundred gallons a year for trips to Gramma's in the other Prius. In early 2017, we got a clean, modern used 2013 Volt for only $13,000.

This 90 percent reduction in fossil fuels and emissions reduced our total $6,000 a year cost of electricity, heating oil and gasoline down to the typical $8 a month minimum electric bill and the $500 or so per year of gasoline for the remaining Prius that we still use on long trips.

Of course, nearly free energy for life from solar requires a substantial initial investment. The way to compare that

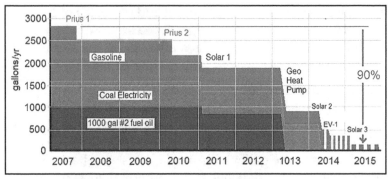

Figure 25. My family's experience getting off fossil fuels over a decade.

investment to monthly utility costs is to amortize the investment over the next 20 years. In our case, this gave a comparable average retail equivalent of about five-cent electricity per kWh, nearly one-third of what we would be paying per month if we had done nothing. This five cents should not just be compared to today's Maryland 14 cent rate, but to the estimated 25 cents per kWh anticipated growth over the same 20 years at an assumed three percent annual rate increase.

### Certainty of Death, Taxes and Utilities

In the past, it was said, "the only things certain in life are death and taxes," but this overlooked utilities that are also inescapably a constant drain on our lives. We all need energy to live. One way to look at the solar investment decision is to consider the impact on your retirement of the monthly utility cost of $100 and what you will have after ten years with $12,000 in a bank earning one per cent interest compared with the same $12,000 invested in roof top solar while paying only the account fee. For the savers who continue to pay $100 per month in electricity bills, at the end of about 11 years, that $12,000 is gone and the savers still face a lifelong burden of $100 a month for electricity for the rest of their life.

Those who turn $12,000 into a solar power investment for life on their roof will retire in security and comfort. Adding another $4,000 to the $12,000 to invest in their own $16,000 solar system will buy about a 6 kW solar system. First they will get back $4800 directly as a credit when they file their federal taxes for a net investment of only $11,200.

Already on day one, they have made an $800 profit, secured their energy for life, eliminated their contribution to fossil fuel emissions, and eliminated $100 per month utility costs for life. They own a physical energy supply system worth $16,000 of equity in their house. After 11 years they have avoided $13,200 in utility bills, have energy for life, and are facing a future of no utility costs for the rest of their years. This makes a net $29,200 value at the end of eleven years from that initial $12,000 investment. It is a much better deal than the saver who has nothing left at the same 11 year point and no security for their energy needs.

The economic analysis presented here was the incentive for writing this book. Since solar is an investement for life, its benefits are valuable for the future retiree, whether single family homeowner with solar panels, or those living in multi-unit dwellings with community solar.

Some might argue that the original $12,000 solar investment depreciates over time and will be worth significantly less at the 11-year point. But I would argue that if it is still producing 80 percent of the original energy at the 25-year point, which is the standard aging rate of solar cell technology, it still retains its original value of providing all your energy to the last day of your life. Even though it has lost 20 percent of its efficiency, one might assume that over 25 years, the load from our more efficient appliances may have also gone down. At present, there is a 30 percent federal income tax credit through 2019 and there may be state and local incentives. These numbers are provided as an example. One should consult a financial advisor for a more refined approach to the analysis.

### Annapolis Friends Meeting

Another example is our Annapolis Friends Meetinghouse. Our small meeting was spending about $1,500 a year on electricity and as much as $4,900 a year on propane. Starting in 2011 we began to explore solar and by January, 2014, a solar leasing company installed a leased 9 kW system (*Figiure 26*).

We fully prepaid $21,000 instead of monthly payments for the next 20 years. The amortized cost per kWh turns out

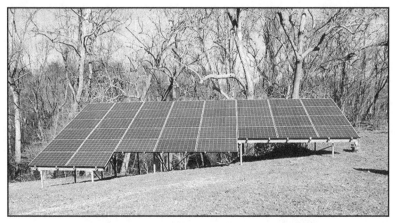
Figure 26. Solar panels at the Annapolis Friends Meetinghouse.

to be about 6 cents per kWh compared to the current 14 cents per kWh in 2015 and the expected 25 cents per kWh after 20 years. This is a huge savings in our Meeting's operating budget. We chose the lease, because non-profits cannot get any of the tax incentives that could save a homeowner nearly 40 percent of the initial investment. But through a lease, the solar leasing company could get all the tax credits and then pass the savings on to us. Outright purchase of our array would have been $53,000, but our prepaid lease only cost us the initial $21,000 and our net monthly electric bills were only the minimum for an account, $8 per month.[50]

What happens in 20 years at the end of the lease? We can extend the lease, start over with a new system when the costs may be lower, or negotiate with the company to leave it in place and avoid the cost of removal and disposal since the resale value of the panels will be negligible at that time compared to new systems then. Already just four years into the 20-year contract, they are offering a complete buyout for around $5000. It should be less as the years go by.

The meeting's other major energy expense and source of carbon emissions was propane for heating. By 2014 propane had nearly doubled to $4 a gallon, so we replaced that fossil fuel system with a heat pump and our total heating cost dropped from $4,000 for propane down to about $1,200 a year in the form of added electricity. The Meeting then approved the purchase of additional solar panels to offset that new energy to bring us back to 100 percent-clean-renewable solar for our meetinghouse.

## Heat Pump Savings

We had planned to go with a geothermal heat pump because of its high efficiency, but on closer analysis we realized that our building is not used like most homes or businesses, but is operated on a significantly skewed schedule. We have no day school or any permanent office staff and the meetinghouse is unoccupied overnight and most daytime hours with most usage occurring in the evening. The thermostat is set back overnight and little heating occurs at night and early in the morning when outside temperatures are the coldest. Most of our heating (except First day) is in the afternoon in preparation for evening meetings. Since the climate of Annapolis has an average outdoor winter temperature at least 40°F in the afternoons, even in January and February, the high efficiency of the geothermal heat pump was not needed. An air-source heat pump would be nearly as efficient at half the initial cost. So we went with an air-source heat pump. The result of switching from propane to electric (to be supplied by solar) not only eliminated our use of fossil fuel and greenhouse gas emissions, but also saved us about $2,500 a year in energy costs and paid for itself in about six years.

## The Charging Sign Initiative

Another initiative that our meeting considered as we transition from petroleum to electric transportation is enhancing the visibility of the ubiquity of electricity and the grid for EVs. The grid is everywhere. Anyone new to EVs may think that public charging infrastructure must be developed before EVs can become a reality, but this is not true. The unique value of EVs is the convenience of EV charging at home or at work while parked and never having to go anywhere and wait for a charge. Since every home and garage and many parking lots already have electricity and all have the same standard 120 volt convenient outlet, in 97 percent of all daily charging needs, simply plugging into a standard 120v outlet is all that is needed for the two-thirds of those in the U.S. who live in single family detached homes. For the other one third, a simple outlet at work can do almost the same thing. In progressive states such as Maryland, new laws encourage multi-family apartments and condos to allow EV charging for residents where practical.

To encourage EV adoption, in 2010 Annapolis Friends meeting put up two simple "EV CHARGING" signs over two of our existing outlets on light poles in the parking lot (*Fig. 27*}. This has been a small beacon of hope and visibility to the green leadings of our community. Several of our members now drive EVs and with the high visibility of our solar system in the front yard, we are seeing more rentals of our building by climate change and energy groups desiring meetings in a 100-percent-renewable-energy-powered facility. To meet this potential growing demand we added another three outlets in 2016 including a level-2 faster 240 volt charging outlet. Until we feel confident, we keep the level-2 charge cord in the office to be used as needed instead of leaving the $300 cord hanging on the EV charging post at the end of the street. In 2016 already we had six percent of our meeting either driving EVs or considering one on their next car transition. Our action to put up a sign at our own meeting house inspired a local initiative that by 2017 involved almost 20 other churches,

Figure 27. Anapolis Friends Meeting EV charging stations.

64

schools, or businesses that had existing outlets and now have signs.[51]

## The Value of EV Charging Signage

The real value of the "EV Charging Outlet" signs is not so much for the one percent that might actually use the charging opportunity but to the other 99 percent of visitors that see the signs and realize that electricity is everywhere. The grid is ubiquitous and since any EV can charge from any standard 120v outlet, there should be no range anxiety if you can plug in while parked.

We even demonstrated a solution to charging for those without driveways that have to park in the street. Bending a piece of conduit to reach out over the sidewalk can legally bring an EV charging cord to a car parked in the street without causing a trip hazard on the sidewalk as shown in Figure 28 (*p.66*). In that photo you can also see in the background our Meeting House's original 9-kW grid-tied solar array that provides all of our electric needs over the year, as well as clean energy for EV charging.

Putting signs over existing outdoor charging outlets is an excellent way for any groups seeking to move forward on clean renewable energy. Owners of outdoor outlets could be approached to see if they are interested in assisting with the transition to clean energy by putting up a charging sign. All they have to do is decide what signage is appropriate for the user of their outlet and if they are going to ask for a contribution to the cost of the electricity. In Maryland, the cost to plug in an EV to 120v is about 20 cents an hour. But for all-day charging, the cost turns out to be about $1 per month per daily miles traveled. For example, a commuter with a 15-mile distance to work would pay about $15 a month for the privilege of plugging in at work every day and paying for the electricity used.

## Workplace Charging Challenge

For employees that drive EVs on a long commute, this charging-while-parked is an everyday benefit and such employees are happy to pay the few dollars a month for the privilege. To facilitate the shift to EVs, the Department

Figure 28. Over the sidewalk EV charging with solar panels in the background.

Figure 29. Volunteer Interfaith BikE Sharing Project (VIBES) at Annapolis Friends Meeting.

of Energy has initiated the Workplace Charging Challenge to encourage employers to support EV charging at work. By 2016 across the U.S. nearly 7,500 charging stations were installed at 757 workplaces, the workers of which own more than 14,000 plug-in EVs.[52]

### Volunteer Interfaith BikE Sharing

Another idea suggested at our own meeting house is the Volunteer Interfaith BikE Sharing project (VIBES, *Fig. 29*). The VIBES concept is that we would keep a few bikes on the Meeting house porch for anyone to use. It works on the honor system. But imagine if this inspired other local churches in our town to also do the same thing. Then we would have bikes scattered everywhere and a system of church volunteers to maintain them. The project idea grew from the observation that at most of our semi-annual Quaker Markets there were bikes donated, and they never sold for more than about $20. So why not keep them on the porch and just let people borrow them for local errands instead of driving with gasoline?

Our parking lot is another asset of the meetinghouse that could be useful in the reduction of fossil fuel burned in our community. Since our lot is usually empty during the day, it offers a place where a distant member can come to town, park for free and then grab a bike to head downtown. There they can shop, visit and tour at will and without the hassle of $2/hour parking everywhere they go or without worrying about the $2/hour bike charge of a commercial bike share system.[53]

## The Transition Movement

In the tenth QIF Focus Book, *Rising to the Challenge: The Transition Movement and People of Faith*,[54] Ruah Swennerfelt explored the exciting growth of the Transition Movement. The summary of the state of solar, wind and electric vehicles in this book will contribute useful ideas to complement that work. With grid-tied solar now cheaper than the utility, solar is an all-inclusive energy source that brings us together in a mutual beneficial web of energy. Instead of seeing the historically fossil-fueled grid as the enemy, we can now see our participation in a cleaner grid through wind and our own grid-tied solar contributions as this new progressive and cleaner energy-web.

The potential revolution that solar and EVs bring to our renewable future is dependent on public education and an awakening to the realization that clean energy and transportation solutions are actually here now and available to anyone. The Transition Movement is well placed to help spread this good news because they have a keen interest in moving forward on community resilience and sustainability to a peaceful, equitable, and environmentally friendly future.

## Chapter Eight
### *What Friends Can Do*

Baltimore Yearly Meeting's Faith and Practice speaks to faith, the life of the Spirit, and using our God-given talents for social and civic responsibility, and peace and nonviolence. How much more peaceful will the world be if we eliminate our dependence on foreign oil and natural gas fracking through energy self-sufficiency powered only by the sun, wind, and other renewables?

As this book goes to print in 2018, the pace of change in energy is so fast that we could not keep up with every new announcement with respect to solar, wind, electric transportation, batteries and energy storage, and the demise of coal. We expect that history will show this as a turning point when the exponential growth of these new technologies began to take off. Rather than try to squeeze in every new press release on the positive momentum toward a clean energy future, this book aims to prepare Friends with all the facts for the new era as the old fossil fuel-powered energies are on their way out and we are on the cusp of very rapid change.

Once readers realize how much they can do in their own lives and how simple, yet significant, these changes are and how much money they can save in the long run, they too will be sensitized to the flood of good news and positive announcements that we are seeing every day.

Over the last eight years, there have been five explosive exponential growth technologies that will change the world and in a very good way.

Solar power became cheaper than coal in 2015 and its growth is exponential.

2) The cost of wind power generation dropped below the cost of coal and gas in 2016 and its growth is exponential. Many countries have reported that 100 percent of their power demand on some days has already been achieved with renewables.

3) The average cost of EVs dropped below the average cost of gasoline- and diesel-powered vehicles in 2013. By 2018 the average cost of the 40 full-sized EVs on the market with incentives was more than $7,000 less than the average cost of fossil fuel cars.

4) Most major car manufacturers around the world have announced that most, if not all, of their future models over the next five to ten years would include electric drive trains. Some countries and large cities have announced total bans on fossil-fueled cars over the next decades.

5) The cost of lithium ion batteries has dropped by almost eight-fold in the last decade. This brings about the magic of the modern electric car, and has major implications for changes in the grid and home energy.

The purpose of this book is to help Friends be prepared for the inevitable major energy decisions they are going to face every few years. Information is given so they will be prepared to make wise choices for a sustainable future. On average a new car is purchased every six years, but each car we buy remains on the road for an average of 18 years before it is finally scrapped. Buying a gasoline-powered car today that would still be on the road long past the climate change tipping point is a very unwise decision. One only gets one chance every 15 years or so to change your home heating system. It would be a travesty to replace an old fossil-fuel burner with a new fossil-fuel burner that will be with you long past the point of no return in climate change.

There are major things Friends can do now:

• Switch from a gasoline car to an EV for local travel or a plug-in hybrid for both local and distance travel.

- Switch from fossil fuel-generated electricity to solar and wind-generated electricity with your local utility.
- Reduce the burning of fossil fuels for heat when a heat pump can do it cleaner, cheaper, and with renewable energy.
- Apply knowledge of these evolving technologies going forward to solve problems.
- Spread the word of these dramatic changes through conversations and study groups.
- Encourage your faith organization (whether Friends Meeting or some other denomination) to educate their congregation and the wider community.
- Emulate Annapolis Friends Meeting and encourage your faith or other organization to update their heating and electricity to renewable energy, and encourage their congregation to buy EVs or plugin hybrids.
- Advocate for mechanisms that make it easier for everyone to go renewable, such as, public and workplace recharging for EVs, community solar, and public utility renewable choices.
- Encourage local car dealerships to offer EVs and plug-in hybrids to their consumers.
- Advocate for policies that support individual solar choices and against those that discourage the use of solar.
- Advocate for renewable energy options for local government operations, such as an EV transport fleet.

An example is the response to the total devastation of Puerto Rico's power grid by Hurricane Maria in September, 2017. All of the island's inhabitants lost power. As this book goes to press five months later, one-third of inhabitants are still without power and expectations are that it will be years before the island is fully restored. But throughout this process of rebuilding, there was some good news. Instead of restoring the old diesel generators, coal-powered centralized power stations, and vast network of transmission lines, they are rebuilding some sites with solar and wind generation in micro-grid systems. Such forward thinking minimizes the potential for future weather-related destruction and replaces the fossil-fuel power with renewable energy sources.

This is the approach we should take everywhere. With the very real threat of global climate change we should not even consider repairing any fossil fuel system, when we can replace it with renewable energy. The technologies are right at our fingertips, and they even save money, reduce risk, and provide security in the long run.

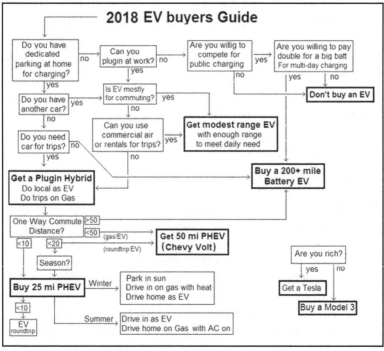

Figure 30. Electric Vehicle Buyers Guide

# *Choosing Your Electric Vehicle*

Although the optimum application of EVs is for local travel and commuting, there are now electric car solutions for almost everyone. The lower cost all-electric models commonly referred to as BEVs (Battery Electric Vehicles), such as the top selling Nissan Leaf, are ideal for daily commuting and local travel, and for families that have more than one car and can dedicate the all-electric car for local use. For single car owners where the car needs to do both local and distant travel, the Plugin Hybrid EVs (PHEVs) are ideal.

### *New Choices in 2018:*

The decision about which EV to buy is a little more flexible but complex in 2018, and accommodates a few more factors as shown in Figure 30 (*page 72*). The key decision points still rely heavily on having a place to park and charge overnight at home. But now EVs with longer ranges open up the possibility for those who cannot charge at home to charge less often, but it does then make them entirely reliant on charging at work or a public charging station.

Prior to the availability of lower cost 200+ mile range EVs, the buyers' guide through about 2016 was simple. If you had a place to park and plug in overnight, get an EV for local use. If you do not have a place to plug in overnight, an EV was not practical at that time. Second, if you only had access to one car and needed to also drive on trips, the logical choice was to get a plugin hybrid which could do both.

But charging always takes much longer than filling a gas tank, so owning an EV and using public charging is not a sustainable concept unless the individual only drives ten or twenty miles a day and could charge once a week at a public station. Charging can take several hours at a standard Level 2 charger. Fast DC public chargers that can furnish up to 80 percent battery capacity in about 30 minutes are becoming more widespread along the major interstates, but as more and more EVs flood the roads, routine public charging will not be sustainable. When thousands of long-range cars need 30 minutes each for high-capacity charging in public, the competition for available spots may be extreme. Also, exclusive repeated routine high speed charging is not good for the long-term life of the battery.

### A Place to Park and Plugin

Since the typical EV needs to be charged every day (typically overnight), the idea of only charging at public charging stations every day for millions of EVs is unsustainable in the long run. The only exception is if you are a full time employee and have a place to plug in at work and your employer allows you to plug in every day, which would cost about 50 cents a day for a 15-mile commute. This charging-at-work situation is the only other eight-hour parking location that is comparable to the overnight charging at home while parked. Two-thirds of those in the U.S. live in single family dwellings and could plug in every night for maximum range the next morning.

### Long Distance Travel

The EV is the ideal match for the 95 percent of our local/commuting transportation needs, but for the other five percent of the time when we have a need to travel long distances (interstate travel), there are many renewable options:

1) Use public transportation by bus, rail or air.

2) Rent a plug-in or gasoline hybrid.

3) Buy a plug-in hybrid that runs on clean electricity locally but can still run on gas for distance.

4) Be a two-car family with an EV for local and commuting, and a plug-in hybrid for trips.

5) Buy a long-range EV with a +200-mile battery.

Making a choice of these options is a very individual consideration. It is uniquely based on the extent of your daily commute and local usage compared to how often you need to go on trips and how far. Also important is your access to another car in the family or how much you are willing to pay the extra cost to have a huge 240-mile battery which is really used for less than 50 miles a day most of the time.

## Models Available in the U.S. in 2017

The sidebar on page 76 shows the plans of U.S. automobile manufacturers for EVs as announced in the media. Figure 32 shows the price, range, and performance (miles per gallon equivalent) for both plug-in hybrids (*top*) and EVs (*bottom*).

## Plugin Hybrids

For those with access to only one car, or who may need to make several hundred mile trips, the plugin-hybrid fully meets that need. The plugin-hybrid not only includes sufficient all-electric daily miles with home overnight charging to meet local transportation needs on 100 percent clean energy, but also has a backup gas engine for long trips at anytime, anywhere. It represents the most flexible transition experience from gas to electric but costs a little more.

Nearly half of the 40 EVs on the market in 2018 are plugin hybrids. These are ideal transition cars as we switch from a fossil fuel economy to an electric one. Plugin hybrids offer a long range similar to gasoline cars, but have a minimum local EV range that allows the car to be used locally, daily, and for commuting entirely on electric and charged at home. Although these models cost more than EVs because they contain both the EV drive train and a backup gas engine with all of its pollution control accoutrements, they completely solve the perception of range anxiety which was keeping many people out of the EV market. Another point in their favor is their popularity with auto manufacturers.

A plug-in hybrid lets them build a similar performance car, but with a smaller gas engine. The battery side can provide the peak acceleration demanded by performance drivers, yet also provide the economy of a smaller battery and smaller gasoline engine while still providing legacy long range. The top sellers in 2016 in the plugin hybrid category were the Chevy Volt and Toyota Prius Plugin. The Volt has a range of 50 miles a day of electric use and has a gas engine that can take over and drive for 300 miles between gas stations. The Prius Plugin has the advantage of a lower cost, but it has about half the battery range of the Volt, but a 600-mile range on gasoline.

The plug-in hybrid is a good transition option, but at some point, the lower cost of the +200-mile EV battery will give comparable range without the added cost and maintenance

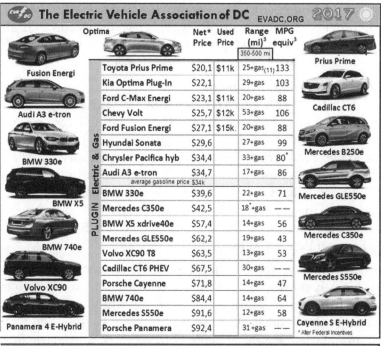

**The Electric Vehicle Association of DC** EVADC.ORG 2017

| PLUGIN Electric & Gas | | Net* Price | Used Price | Range (mi)[3] | MPG equiv[3] |
|---|---|---|---|---|---|
| | Toyota Prius Prime | $20,1 | $11k | 25+gas(11) | 133 |
| | Kia Optima Plug-In | $22,1 | | 29+gas | 103 |
| | Ford C-Max Energi | $23,1 | $11k | 20+gas | 88 |
| | Chevy Volt | $25,7 | $12k | 53+gas | 106 |
| | Ford Fusion Energi | $27,1 | $15k. | 20+gas | 88 |
| | Hyundai Sonata | $29,6 | | 27+gas | 99 |
| | Chrysler Pacifica hyb | $34,4 | | 33+gas | 80* |
| | Audi A3 e-tron | $34,7 | | 17+gas | 86 |
| | average gasoline price $34k | | | | |
| | BMW 330e | $39,6 | | 22+gas | 71 |
| | Mercedes C350e | $42,5 | | 18*+gas | —— |
| | BMW X5 xdrive40e | $57,4 | | 14+gas | 56 |
| | Mercedes GLE550e | $62,2 | | 19+gas | 43 |
| | Volvo XC90 T8 | $63,5 | | 13+gas | 53 |
| | Cadillac CT6 PHEV | $67,5 | | 30+gas | —— |
| | Porsche Cayenne | $71,8 | | 14+gas | 47 |
| | BMW 740e | $84,4 | | 14+gas | 64 |
| | Mercedes S550e | $91,6 | | 12+gas | 58 |
| | Porsche Panamera | $92,4 | | 31+gas | —— |

Labels: Fusion Energi, Audi A3 e-tron, BMW 330e, BMW X5, BMW 740e, Volvo XC90, Panamera 4 E-Hybrid, Optima, Prius Prime, Cadillac CT6, Mercedes B250e, Mercedes GLE550e, Mercedes C350e, Mercedes S550e, Cayenne S E-Hybrid

\* After Federal Incentives

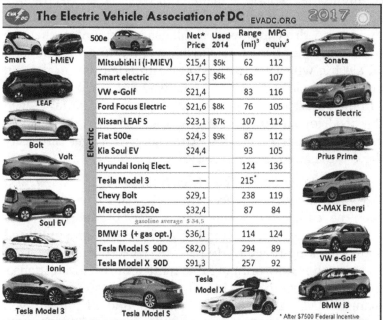

**The Electric Vehicle Association of DC** EVADC.ORG 2017

| Electric | | Net* Price | Used 2014 | Range (mi)[3] | MPG equiv[3] |
|---|---|---|---|---|---|
| | Mitsubishi i (i-MiEV) | $15,4 | $5k | 62 | 112 |
| | Smart electric | $17,5 | $6k | 68 | 107 |
| | VW e-Golf | $21,4 | | 83 | 116 |
| | Ford Focus Electric | $21,6 | $8k | 76 | 105 |
| | Nissan LEAF S | $23,1 | $7k | 107 | 112 |
| | Fiat 500e | $24,3 | $9k | 87 | 112 |
| | Kia Soul EV | $24,4 | | 93 | 105 |
| | Hyundai Ioniq Elect. | —— | | 124 | 136 |
| | Tesla Model 3 | —— | | 215* | —— |
| | Chevy Bolt | $29,1 | | 238 | 119 |
| | Mercedes B250e | $32,4 | | 87 | 84 |
| | gasoline average $ 34.5 | | | | |
| | BMW i3 (+ gas opt.) | $36,1 | | 114 | 124 |
| | Tesla Model S 90D | $82,0 | | 294 | 89 |
| | Tesla Model X 90D | $91,3 | | 257 | 92 |

Labels: Smart, i-MiEV, LEAF, Bolt, Volt, Soul EV, Ioniq, Tesla Model 3, Tesla Model S, Tesla Model X, 500e, Sonata, Focus Electric, Prius Prime, C-MAX Energi, VW e-Golf, BMW i3

\* After $7500 Federal Incentive

Figure 31. Plug-in Hybrid models available in the U.S. in 2017 (*top*) and EV models available in the U.S. in 2017 (*bottom*)[.66]

of the small gas engine, gas tank, exhaust system, catalytic converter, pollution controls, complex transmission and all those hundreds of moving parts required in a hybrid. By 2018 the 200+ EVs were at cost parity with gasoline cars.

### Choosing a Plugin Hybrid:

The idea of a plugin hybrid is to buy one with enough range to do most of your local, daily and commuting trips entirely on battery while also having the gas engine for backup and for longer range trips. Do not make a choice based on mixing electric and gas miles in the same trip routinely. This is a waste of the EV battery and a higher overall emissions scenario because if the gas engine is used only for the tail end of every commute or short trip it is operating cold, inefficiently, with poor gas mileage and much higher emissions.

Instead, the goal of the PHEV is to do most local trips exclusively on electric and any longer trips exclusively on gas when the engine can operate more efficiently and can save wear on the battery. A good example is a 20-mile one way commute that cannot be made both ways on all electric in a 25-mile PHEV, but it can make it one way entirely on battery. In this case, drive into work on all gas in the winter when the waste heat is of benefit to staying warm and drive home on all electric when the car has been sitting all day in the sun and requires little EV heat on the way home. Conversely, in summer, drive into work exclusively on the EV battery when the need for air conditioning is less, and then drive home exclusively on gas when the extra energy from gas can easily drive the car and AC efficiently.

The isolated box in the lower right of Figure 30 (*Are you rich?*) is a tongue-and-cheek tip of the hat to Tesla that had been the leader in EVs and prior to 2017 produced the only available long-range +200-mile EVs. Tesla had long promised a low-cost long-range EV to meet the $35,000 average cost of all cars and is now bringing it to market in 2018. In response, as this book goes to press in early 2018, we are seeing the other manufacturers also racing to that goal and Chevrolet even beat Tesla to market by several months with the 230-mile Bolt.

## Cost of EVs

Battery BEV's have the potential to cost much less than comparable gas and hybrid cars because they do not need all the duplicative gas engine and pollution controls. But the goal of the car companies is to make a profit. And with the average cost of all cars being about $35,000 in 2017, the car companies would rather keep adding features and keep the EV cost up than to sell more lower cost, lower profit cars. Since the public perception that longer range is a desired commodity, all manufacturers were in a race to meet this customer-perceived market with new +200-mile EVs that cost the same $35,000 average price as gasoline cars and yet compete nearly equally with them on range. By 2017 the price with incentives of the 100-mile EVs were quite low as manufacturers were clearing inventory to make room for the coming wave of +200-mile EVs from Tesla, Chevy, Nissan and others. But for the commuting-only or second car, there will always be a market for shorter range BEVs.

At the end of the 2017 model year, Nissan was selling the new 2017 Leafs for as low as $15,000 with incentives, but this was mostly to clear out room for their coming 2018 model with nearly double the range. Several global manufacturers have already announced that all of their car models will have electric motors in them either in the next few years or by about 2025. Electric drive is becoming mainstream. Some will be pure battery electric and some will be plug-in hybrids, but clearly electric transportation at least for local travel is the wave of the future, if not the present.

## Buying a Used EV

There will always be a used market for shorter range EVs. They might be ones that were originally manufactured with shorter ranges or ten-year-old long-range EVs that have lost some of their initial range but still have plenty of overkill for the local commute. In 2017, the used price for three-year-old Leafs was under $9,000 and for Smart EVs under $6,000. Used Chevy Volt plugins were going for under $14,000.

Some people point to this drastic depreciation in price as a problem with EVs, but $7,500 of that price reduction is due to the Federal tax incentive. Only the new buyer could take the $7,500 credit, so this instantly devalues the resale value by the same amount since the second-hand customer cannot take advantage of it. Another reason for the lower resale value is the ever-present perception of the need for maximum range whether needed or not.

## Batteries in used EVs

After eight years of EVs in production and well over ten billion total miles on the road, the general consensus is that hardly any EV has actually worn out its battery beyond initial predictions in normal use. There was some early unexpected loss of range for some Leafs in 120°F temperatures in Arizona, but these problems were corrected. There are now many EVs still on their original batteries with well over 150,000 miles and going strong. Further, the price of replacement batteries continues to fall. For example, the original Prius $6,000 hybrid battery from 2004 was down to well under $3,000 (dealer installed) a decade later. And there were tens of thousands of used batteries available on the salvage market for under $1,000. A robust after-market industry has evolved that is currently selling fully reconditioned Prius batteries with five-year warrantees for about $1600 because hybrids will need them after about 10 years.

But the Prius battery is small. For BEV's with a battery 20 times larger than the Prius, a better example is the 100-mile Nissan Leaf battery which Nissan guaranteed could always be replaced for $5000.

Battery EV manufacturers guarantee at least 80 percent range over the 100,000 to 150,000-mile life of the car, so there is still plenty of mileage left for the shorter range commuter for years to come.

# Appendix B
## Choosing Your Solar

The purpose of this section is to lend a hand to homeowners considering buying a solar system, a substantial financial investment for their future and retirement. In researching solar options, the homeowners may encounter a myriad of sales options and purchasing plans, so this unbiased guidance, not associated with any sales pitch, is provided to help avoid some common overreaching sales claims and misguided sales information. In most cases the homeowner will be installing on their own property, but in some progressive states, local utilities may also offer virtual net-metering. In this case, the solar generation credits on that electric meter at the solar production property can be applied to any other property or electric meter in the same utility area.

### Roof Age

How good is the roof and does it make sense to cover it with solar panels for the next two decades? Some installers will suggest that an old roof needs to first be renewed before making the major solar investment. In many cases a new roof may cost as much as or even more than the solar system. But if the roof is in relatively good shape and not leaking, then covering it with solar panels can substantially extend the working life of that roof because the solar panels protect the roof from the harsh elements, hot baking sun, hail, rain, snow, ice, and wind. The panels are not water tight because there is a quarter-inch gap on all edges. The roof still gets wet, but is not pounded by the rain.

An alternative is to place your panels in your yard on a ground mount, which eliminates the roof as an issue and gives greater flexibility in orientation and placement. It might cost a little more for a supporting structure.

### Solar Roofing Materials

One evolving consideration in the solar roof decision is the recent rise of solar shingles. New building codes are being proposed that would require new buildings to be "solar-ready" such that the wiring for a solar roof would already be in place.

The current costs for roofing installations range from about $6 per square foot (sqft) for asphalt, $14/sqft for tile or metal, and $16/sqft for slate. The 2018 estimate for Tesla solar shingles is about $40/sqft, but since only one-third of the total roof will actually be producing solar power, Tesla assumed only one-third of the roof would be solar shingles. For a modest house with a modest solar size system, the cost would be $22/sqft. This is competitive with the cost of adding solar panels which currently cost about $2.75 per Watt installed or about $39/sqft for the actual silicon-producing area. But the power produced from any of these roof solar systems over 20 years more than pays for the roof. This is the magic of solar and free power for decades once the system is installed.

### Pointing Direction of Solar Panels

Historically for off-grid systems, solar panels were all pointed south for optimum power production, but that was before the introduction of grid-tie system economics. For independent off-grid battery-only systems, the panel must face south and be pointed at the lowest angle of the sun in winter on the shortest day of the year (21 December) or the system will not collect enough energy for that worse case day's operation.

But with solar systems that are tied into the electrical grid, the orientation of solar panels is not that much of an issue. The important issue is the total energy collected by the array over the entire year, not the shortest winter day. For almost any direction, there is a time of year when any given orientation of a solar panel can produce equal or more power than the ideal southern array at that same time of year. Panels are now added to any sunny roof with little regard to the exact

direction so long as it faces relatively clear sky soʳ
between east, south or west.

Pointing the array due east or west results in only a 15 percent loss of annual production compared to the ideal array because in the summer the sun actually rises in the northeast and sets in the northwest. These east and west arrays produce more power in the summer than the ideal southern array produces which is shaded in the morning and late afternoon. A roof with both east and west-facing sides lets that homeowner produce twice the power of just one roof and, instead of 85 percent, it is 170 percent of what the same panels in one southern array can produce. For those who are interested, one can use the National Renewable Energy Laboratory's on-line program called "PVwatts" <pvwatts.nrel.gov>, which calculates the total power that will be generated from any combination of azimuth pointing angle and tilt. This website is the gold standard for most solar designers and one can assume it, or a similar on-line calculator will remain available.

Since solar panels are so cheap these days, many homes with small roof areas are even considering panels on gentle sloping north, northwest and northeast facing roofs, as long as they are unshaded. These panels produce practically nothing under the low winter southern sun, but during the summer when the days are twice as long and the sun is 47 degrees higher in the sky, they will produce additional power. Depending on the latitude, they may produce 60 to 68 percent of what the ideal southern array can produce, but that additional source of solar power could be cost effective, depending on the need for more array power and the cost of installation.

### Array Tilt Angle

When determining the design of our Annapolis Friends Meeting we looked at the variation of power production relative to the tilt angle of the array. Because the array is mounted in our front yard and is in a residential neighborhood, we wanted to minimize the visual impact of the array. We ran the PVwatts model multiple times at the various tilt angles and plotted the results in Figure 32 (*next page*). For

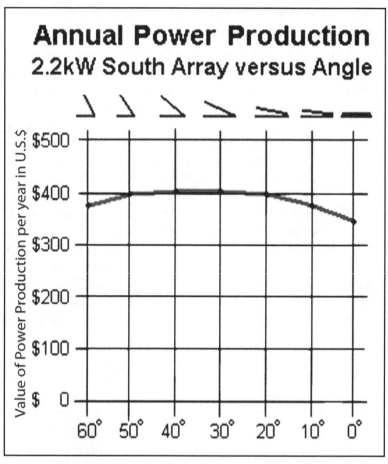

Figure 32. Effect of the tilt angle on solar power production of my solar array in Maryland. These calculations will vary for different locations and facing direction.

any angle between 20 and 50 degrees, the annual power production varies only about one percent, so even though the ideal maximum power angle was 40 degrees, we were able to reduce the visual obstruction by half to only 20 degrees and still produce 99 percent of maximum annual power. Any angle below 15 degrees is not recommended by installers because there is inadequate rain runoff to wash the panels clear of dust and debris, and snow will not fall off in winter. As an extreme example of the flexibililty in azimuth pointing, solar panels lying flat on the ground will still produce 80 percent of what the ideal southern facing array can produce. Being flat, they are facing all directions at once. But the lack of self-cleaning in the rain cancels any advantage being flat might have.

## Lease or Buy

Choosing how to pay for the array is an important consideration for most people. Like any other investment, you get the maximum benefit when investing your own money so no finance charges are involved. But a lease is a good way to stop using fossil-fuel-powered electricity immediately and convert to clean renewable solar energy with zero initial investment. In this case the solar company is investing in the solar production on your house and, instead of paying the power company for electricity as in the past, you will pay the solar company a monthly repayment on their investment on your roof. Usually the solar lease company will guarantee that your monthly payments to them will be around 10 percent less than what you used to pay for the same electricity from the utility. Assuming you are still tied to the grid, and you still have a utility bill, 90 percent of what you pay is to the solar finance company. There is always a remaining flat fee just to have the utility account because that is where you store all your energy during the day and in the summer for use at night and in the winter. In Maryland that flat rate is about $8 per month and is a bargain compared to having $5,000 worth of battery storage.

If you decide to lease, look at the lease calculation numbers carefully. To make sure that your solar lease payment is less than what you used to pay for the same energy, the solar company makes some assumptions. They assume that the annual cost of electricity from the utility will rise. A common figure used is three percent per year, so they include that rise in your financing payment schedule. This shows the rising value of your solar-produced power, but it means that what you pay to the leasing company is built to rise by that same percentage over the 20-year life of the lease. Originally that was a good assumption. In 2008 at the beginning of the solar revolution the utilities and all analysts were in agreement that electric rates would continue to rise to meet rising demand and the need to build more coal and gas plants.

But demand has not increased at the rates predicted in many areas. The death of coal, the low cost of solar and wind, and higher efficiency appliances and lighting of the last decade has actually kept electric rates almost flat or even going down somewhat. In those cases the three-percent escalation factor did not happen and some people who leased solar installations might not be seeing the full savings promised.

## Tax Incentives

One of the key elements of investing in solar is the ability to take the 30 percent U.S. federal tax credit, plus any state, country, or city credits. This only applies to those who purchase their array outright and only to people who pay taxes. For our nonprofit church, none of these tax credits would apply if we purchased the array directly. But by leasing, the leasing company was able to take all of these investment credits up front and pass them along to the church in lower overall lease terms. Although our meeting chose to lease due to these advantages, we did not want to pay 20 years of finance leasing charges to the solar company so we made a single up-front payment for the entire 20-year lease. By avoiding 20 years of finance charges and escalating rate estimates, that payment was about half of what a full purchase would have cost.

## End of Lease Negotiations

What happens at the end of the lease is still open for negotiation since none of the 20-year leases initiated during the rise of solar starting around 2010 have run their course as this book goes to print in 2018. For our Meeting's lease, which cost us a single up-front one-time lease payment of about $21,000, the solar company also offered to sell us the array at the time of the initiation of the lease for about $6,000. They included a declining buy-out price table that decreased to about $4,000 after 20 years. Given that the system would still be producing over 80 percent of its initial power rating at that 20-year point, this is still a bargain. But we are holding off. At the rate of price decline, our old panels would have practically no-resale value for the leasing company at that time compared to 20 years of advancement of solar panels, so we think the value of the array to the solar company at that time will be less than the cost of labor to remove it. We are waiting to negotiate that at a later time and offer them the choice to leave it in place at no cost to either of us and thus they avoid their cost of removal.

### Purchase

For the homeowner that has been saving for retirement and their future life in their home, outright purchase of the array is the best investment. Generally the homeowner will see at least a 10-percent annual rate of return on her investment over the life of the array. The real value of that return rises over time when compared to future rises in the cost of electricity. These percentages of return on investment cannot be met by banks, or stocks or bonds at least not with the same sense of security. Your investment in banks, stocks, or bonds are just numbers on a piece of paper. But your investment in solar is actual hardware equity on the roof of your own house that you own and no one can lose it through mismanagement or corruption.

### Choosing the Inverter System

My first panels in 2012 were 220 Watts, but today the output of a typical 3.3 by 5.4 foot solar panel in 2017 is about 270 Watts. This equates to about 30 volts at 9 amps, so this must be converted to the 120/240 volt AC grid in your house. There are three ways to do this. 1) String inverters involve connecting "strings" of solar panels in series to a central inverter that is usually mounted in the basement or along the side of the house. 2) Micro-converters are small boxes containing the conversion electronics attached to the back of every individual solar panel so that the power produced by each panel is fed to a cable carrying the 240 VAC directly to the home's distribution panel. 3) Optimizers are a combination of the two that consist of small electronics boxes mounted on the back of every panel, but most of the critical grid-interface electronics is on a wall somewhere like a string inverter. Installers have their own preferred approaches for each specific installation, but any one does not fit all cases and it is useful to understand the differences.

## String Inverters

The original method was to connect the solar panels in strings of ten to fourteen panels in series, each producing 30 volts, and then connect the summed series of 300 to 450 volts direct current (DC) to a single string inverter that converts the DC to the 240-volt alternating current (VAC) to feed your house and the grid. The advantages of string inverters are simplicity, ease of maintenance of having the inverter in a protected easy-to-reach location, and lower copper wire costs. The amount of copper wire needed to bring solar power to your power distribution box is proportional to the square of the current. If ten panels were all connected in parallel, wire capable of ten times the current or 100 times the amount of copper is needed to safely carry the same current, though the voltage would only be 30 volts. If the same ten panels are connected in series, the voltage goes up to 300 volts, but the 9 amp current remains the same. With the series wiring in string inverters, the whole array is only operating at 9 amps, and so the array can be connected with inexpensive common #14 or #12 house wire. When the array is a long distance to the roof or is ground mounted some distance away from the distribution panel, this can be a significant consideration in wire costs. The only disadvantage of series string arrays is that all ten panels must be mounted in a single direction so that they all produce the same current under the same solar illumination.

Another consideration is when you opt for string inverters and have more than 14 panels, you have the choice of one big string inverter for the multiple strings (in groupings of ten to 14 modules for each string) or for multiple string inverters connected in parallel. I went with multiple smaller string inverters because my arrays grew over time and because I like having three identical separate inverters for my three strings to give me redundancy rather than having all arrays going through a single possible point of failure. It cost me about \$300 more to go with separate inverters, but I liked the redundancy. If one failed, at least the other two continue to produce while waiting for repair of the third.

### Microinverters

One advantage of microinverters is that they are easier to install than string inverters. Since each panel has its own electronics, one can get a report on the daily and hourly performance of each individual panel. Microinverters are an advantage on disjointed roofs where it is not possible to congregate at least ten or more panels into one-direction string arrays. With microinverters each panel can point at a different tilt and direction and still contribute full power when it is illuminated independent of what adjacent panels produce.

The disadvantage is that they cost more. Since the cost of panels is now one-tenth of their previous cost and the cost of microinverters has not gone down as much, using them adds proportionally more to the total cost. Microinverters made some sense back in 2008 when a single solar panel cost \$1200 and micro inverters only cost \$250. But now that panel prices are below \$150 each, the microinverter can cost more than the solar panels.

Besides the cost, another disadvantage of micro-inverters is that they are mounted on the roof under the solar panels where they see the full -10°F to +160°F temperature range. This is not good for electronics. Although the microinverter's individual reporting can identify a failure to a single panel module, having all this electronics on the roof makes repair and replacement labor intensive compared to repairing a single string inverter on a wall.

### Optimizers:

Because of the declining economics for microinverters, the modern equivalents are lower cost optimizers, which are small electronics boxes still mounted on the back of every panel but with most of the electronics in a box mounted on the wall. The optimizer module can report on the condition of each panel and can adjust the output under a variety of lighting conditions.

Some claim that this approach has the combined benefits of both the microinverters and string inverters, but others, myself included, see that this is adding all the vulnerability of roof mounted electronics to a wall unit and therefore is no net advantage over a string inverter. Another disadvantage is that optimizers on the roof can act like antennas and generate radio interference for anyone listening to a nearby AM radio. The solution requires placing clip-on filters on every one of the optimizer modules on every one of the panels, both their inputs and outputs. In comparison, if noise is generated in a single string inverter, that is more easily solved by putting a clip-on filter on the single unit in the basement.

### Shading Considerations

A big consideration in choosing between central string inverters and microinverters or optimizers is the issue of shading. Each panel consists of about 60 cells producing half a volt that are connected in series. A series connection is like a "chain and weakest link." If there is shade, such as one leaf covering half of a single cell that reduces that cell's production of current in half, then the entire output of the entire string is reduced by half. Fortunately, every panel is segmented into three separate series strings of only 20 cells and each string is bypassed by its own bypass diode, so in the case of any shading on a cell, only the power of that third of the panel is lost and has no other impact on any segments or other panels no matter what kind of inverter is used.

Many microninverter and optimizer salesmen use the weakest link analogy to say that the entire 12 panels, all in series, are all impacted at the slightest shade, and that is why you should buy the more expensive microinverters or optimizer system. But because of the one-third segmentation and bypass diodes, this is not true. The third of that panel is going to be lost even if that panel has its own microinverter or optimizer. Even though they are connected, each panel is still the same string of three segments with the same shading losses and bypass correction.

Shading impacts the choice of inverters if they receive modest shade all over the array at the same time. For example, with winter sun coming through trees, most of the panels affected are operating under reduced conditions and the microinverters or optimizers can at least still produce power for those few panels that are still getting full sun.

That is why large roof arrays where all the panels point the same way with no mid-day shading are usually wired to a string array for lower cost and simplicity. Microinverters and optimizers are only used when there is no room for a large single array, when panels have to be placed at odd angles and different orientations and where there might be significant shading due to obstructions. Even large arrays in treed suburbia will have some shading at the east end in the morning and the west end in the afternoon. But neither lasts long. Most of the daily solar production at the lower winter sun angles is during the middle six or so hours of the day. The shading in the morning and evening occurs at such poor angles that the loss of a few percent is insignificant.

### Maintenance

Solar panels usually do not require maintenance. With a leased array, the solar company is responsible for all maintenance needed, if any. This is an advantage unless the 20-year finance package has been sold by the solar company to another finance company. A direct purchase solar system is usually installed under at least a one year warranty. If the installation shows any problems during the warranty period, the installer is happy to troubleshoot and make repairs. But once early installation problems are corrected, solar systems are very reliable since there is nothing to wear out or replace. The only risks include heavy hail bigger than golf balls and direct lightning strikes, both of which should be covered by homeowner insurance, or inverter failure. In the case of inverter failure, the cost of a typical 4 kW string inverter is now less than $2000.

All three systems have the option of a WiFi module to report your array performance on an hourly, daily, and annual basis to your PC or smart phone. If one module out of an array fails, the system will simply bypass that panel. In the case of a string array, the individual panel is not reported, but its loss is observable because the overall array output will then be less than for a similar day with all panels operating. A test with a voltmeter can locate a bad panel, though this is more difficult if the array is on the roof.

## APPENDIX C

## *Choosing your Backup and Emergency Power*

Throughout the Chapter 4 introduction to home solar we indicated that batteries have nothing to do with the highly economical recent rise in the value of solar for the home. If one has access to the grid, batteries for daily solar storage with their significant cost, lifelong maintenance, and periodic replacements, are neither needed nor economical. Chapter 6 argued that the costs of home energy storage are coming down rapidly, but storage is still economical in most areas.

The question always arises from homeowners considering solar, "what about when the grid goes down?" The answer is very simple, "Do whatever you do now." Adding grid-tied solar as a source of economical lower-cost, emissions-free power does not change anything with respect to what you do when the grid goes down at your house. What you do depends entirely on your unique and individual situation and the rapidly changing future of battery storage.

I am reminded of a family who retired to their dream home, added solar and included a $10,000 whole house battery backup system. After many years the big four-day power outage finally came along, and they were happpy living normally with their whole house back up system... for about 8 hours. Then the batteries were dead. After an irate call to the installer, they discovered that the batteries are only good for about five years before they all have to be replaced.

In Maryland, our power company claims about a 99.96 percent reliability which means on average we lose power about four hours a year. At twelve cents per kWh and given that the average home needs on average about one kW of power, this is about 50 cents worth of lost electrical power per year. How much are you willing to invest in a backup power system to meet that 50 cent annual need?

Although batteries do not add any economical value as a part of a modern grid-tied, net-metered home solar system, there may be a small place for them as part of a minimal backup system for those rare emergency grid-outages. This section discusses some ways to provide emergency backup power with and without batteries.

### Assessment

The first thing you need to consider about a backup system is how much emergency power do you really need, to what level of comfort, and for how long? Even if you have someone on a respirator that makes it a life or death need for emergency power, it is far more economical to have a small Uninterrubtible Power Supply (UPS) backup just for the respirator and not one a hundred times larger for the whole house. Backup power is a very personal, individual consideration.

### Emergency Lighting

With modern LED bulbs now, candles and propane lanterns are simply obsolete. A single candle consumes about 50W of carbon energy just to put out a feeble one Watt of light. A propane or gas lantern consumes about 1000 Watts of carbon energy to put out the equivalent of a 10-Watt light. And both are dependent on a continued source of fossil fuel that might be hard to find in an extended emergency. But a simple LED lantern or flashlight can be charged every day with a small solar cell and used every night. Be sure to have some on hand.

LED bulbs now are also a great advantage in the home and drastically reduce the power needed for emergency lighting. My home has about 50 light bulbs. If the kids were home and all of the bulbs were incandescent and were lit, we'd need at least 5,000 Watts for backup lighting power. But during a power outage and being conservative maybe only ten bulbs in several rooms are needed at a time and this is now only about 50 Watts with LEDs, or about one percent of what we used to use!

The problem is how to get power distributed to them all from your emergency power source. There are three legal methods.

First is to have enough extension cords to run to a lamp in the main rooms that need emergency lighting. The second is to have a whole house generator transfer switch installed by a licensed electrician costing several hundred dollars. The third is to have the electrician wire all your light circuits unconventionally to a simple A/B switch where side A goes to the existing circuit breaker and side B goes to a standard 120v plug that you can run to a generator. On this circuit it is also advisable to put the refrigerator since that is usually the only other home-critical load and it is small enough to remain under the standard 15 amp circuit.

### Refrigerator/Freezers

An important consideration even before lighting during any extended outage might be the refrigerator and freezer. The name-plate rating may indicate that the refrigerator needs 800W of electrical current, but this includes the defrost heaters that will not come on immediately in an outage. We can assume that without the defrost heaters the refrigerator will run on about 250 Watts. To power this you still need at least a 1500-Watt surge power source, because the refrigerator compressor draws up to ten times the average current for a fraction of a second each time it starts. Between the lighting (about 50 Watts) and the refrigerator (about 250 Watts when running) plus surge power, a 1500-Watt source of emergency power is a good choice to meet emergency needs and occasional peak loads during a long term power outage.

### Backup Generators

An efficient small generator can generate power from gasoline for about four kWh per gallon of gas, or about 75 cents per kWh (five times the average utility cost). The purchase price for a 1500-Watt generator is about $400 plus the $15 can of gas (5 gallons) to run it. If the average power you need is about 500 Watts for 24 hours a day, you would need about three gallons per day. A new class of generator called "inverter generators" are more efficient and quieter, especially those pioneered by Honda, but they cost about twice as much as the cheapest generators. Still, the quiet and economy of these generators is well worth the added cost.

## Generator Reliability and Testing

In the five years since I bought an emergency backup generator, the power has never gone out longer than the time it took me to get off the couch, find a flash light, dig out the generator, find the gas can, and start the generator. Since my generator was five years old and had never been used, I decided to crank it up in anticipation of a coming storm, but it would not crank. Fortunately, we never lost power, but in the rain storm, I put up a tent over it and worked on it every night for a week before I finally got it to run. To be sure to keep it viable for future outages, the prepared owner would have to keep fresh gas in the gas can, a stabilizing chemical in the tank, and run it for an hour or so every month. But if the point is to reduce fossil fuel emissions, it is an anathema to run a gas generator for an hour every month just to keep it ready for a four-hour outage once a year. Even worse is the fact that such small portable generator engines do not have any of the modern catalytic converters and other exhaust recovery systems now required in all automobiles, making that one hour "test" per month the same as a week of average commuting in a modern gas car!

## Hybrid or EV Power

But better than generators, an efficient back-up option is to use your hybrid or EV if you have added an inexpensive 12-volt inverter to your trunk or under the hood. Almost all EVs and hybrids still use a big 12-volt battery for all of the usual automobile accessories. The use of Hybrids or EVs for readily available backup power is still an untapped and unmarketed resource potential from the manufacturers.

The Nissan all electric Leaf EV even has a home interface (available only in Japan so far) that allows the house to draw power from the EV car battery. If the battery is a 32 kWh battery and the 24 hour average draw of a home during emergency outage conditions averages around 200 Watts, then the EV battery can run the house about a week before the EV must be recharged from some other source of power, such as a neighbors outlet or one's own solar.

For the hybrid, it is similar. The powertrain in any hybrid contains a very efficient gas engine driving at least a 50,000-Watt generator to drive the electric motors that drive the car. Compared to a portable generator burning gasoline, the hybrid engine has extremely low emissions because it has a catalytic converter and meets all of the modern low emissions federal guidelines. Your hybrid is used almost every day, so you know it works.

This 50 kW generator in a hybrid is enough to power dozens of homes (running full throttle), but so far, the auto manufacturers have not capitalized on this vast potential nor developed a safe way to provide this power to the general consumer. Both Toyota and Nissan have demonstrated whole-house power from cars to test markets in Japan, but so far have not offered it in the USA.

But there is an easy "do-it-yourself" approach. Just add an inexpensive 12-volt to 120-VAC inverter with 1500-Watt capacity to the hybrid as shown in Figure 33. A Toyota Prius with about eight gallons of gas in the tank, and providing about the same 200-Watt average power through the inverter can run for almost a week because the hybrid will only run

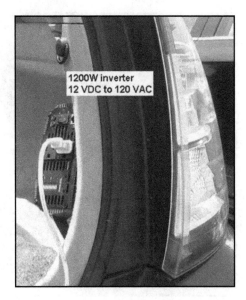

Figure 33. Using a hybrid for back-up.

the gasoline engine when it is needed to recharge the battery. Meanwhile the emergency power is drawn from the 12-volt battery which is maintained by up to 100 amps from the hybrid system battery as needed. Although the Prius wastes about 200 Watts just being powered up, it can stay in this standby condition providing continuous power for as long as the gasoline lasts, only starting the engine as needed to maintain the battery. The car just takes care of itself.

For a Battery EV, Figure 34 shows a one-kilowatt pure sine inverter installed under the hood of a Nissan Leaf. The device plugged into it in the photo is a handy "killawatt" device that can read and analyze power for the curious. It is best to get an inverter that produces pure sine waves because the cheapest 12-volt inverters called "modified sine wave" produce "chunky" waveforms. They are electrically noisy and some electronics systems may not operate properly with them. Inverters that produce pure sine waves giving power indistinguishable from the power company cost about double the modified sine ones or about $0.20 per kWh, so a 1500 kWh inverter would cost around $300. I have never personally had

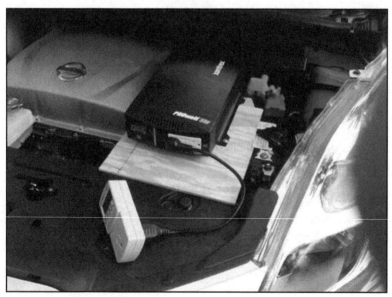

Figure 34 shows an inverter installed in a Nissan Leaf.

98

anything not work on the cheaper modified sine inverters, so I use them exclusively and keep a pure-sine inverter as a backup.

### Value of an EV during Power Outages

People often overlook how vulnerable the gasoline/diesel powered vehicle is during an extended power outage. But after standing in line to get gas when the gas stations have either run out of gas or don't have any electricity to pump it is something they will hardly forget. Even in the aftermath of hurricane Sandy in 2015 with millions of people without power in New Jersey and Long Island, satellite photos revealed that about half of the neighborhoods still had electricity. In such a situation, the glass is still half full for the EV driver who only has to drive to a nearby neighborhood where the lights are still on and ask to plug in to any available 120-volt outlet, the cost of which is only 20 cents an hour to recharge.

However, most of the public and neighboring homeowners still do not understand how inexpensive electricity is for an EV. Although it is only about $2 for an overnight 10-hour charge, the uninformed public sees something quite different. Since they may pay $30 for a fill-up for their small gas car, they are quite leery of someone asking to plugin for an overnight fill-up charge and find it hard to believe that it only costs about $2 a day to help out their neighbor with an EV. The true value and low cost of fuel for the EV should be understood by everyone. An EV can charge anywhere, anytime, from any standard 120v outlet for a USA average 20 cents an hour, even better if they can plug into their own unlimited clean home solar every day.

### Home Daytime Solar Power Backup

Anyone with home solar has thousands of watts of power available during the day. And even when the sky is overcast, the typical array is producing at least 10 percent of rated power. For a 10-kilowatt solar system, this can be about 1000 Watts on cloudy days, enough for most backup critical systems. The problem is that the grid-tie system stops delivering that power when the grid goes out.

.t the good news at more recent grid-tie-inverters are including a 15 amp backup 120 VAC power outlet built into the inverter as shown in Figure 35. This inverter AC outlet power has no battery backup, but does provide about 1500 Watts of emergency 60 Hz AC

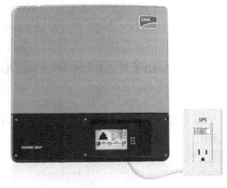

Figure 35. Inverter AC outlet power

power via that outlet when the grid goes down and there is still sun. This 1500 Watts at 120v 60 Hz is enough to operate any conventional 120-volt appliance including charging any EV through its standard charge cord.

Anyone thinking of investing in a home grid-tied solar system should get an inverter (typically a string inverter) with this backup outlet. The added cost of this feature is only a hundred dollars, compared to other models without it. You can only use this backup outlet during the day while the sun is shining. At night is when you have to consider other forms of backup power.

### Batteries with Solar

As we have mentioned before, it is impractical to try to store daily solar power in huge battery banks when we can store it in the grid for free. In contrast, an off-grid isolated system requiring a full day's energy storage, demands that nearly $2 of every $3 invested would be in the battery storage system and life-long maintenance, and only one-third of the solar system investment is actually generating power. That should convince anyone that current battery storage has no place in a modern economical solar power system that has access to the grid.

However, when the grid goes down, we still might need some night time power to run the refrigerator and some lights over an extended outage. The first choice is to simply plug into the 12-volt inverter you have added to your hybrid or EV as described in the earlier paragraphs (cost about $150 to $300). But that ties the car to the house during darkness and leads to inconvenience during long power outages. So it is nice for solar homes to have a minimum battery backup.

For our home, we use two marine 12-volt batteries (cost about $200) wired in parallel to drive a typical auto 1500-Watt inverter ($150) plus a common 12-volt battery charger ($50) to charge them during the day. This charge is either from the grid when it is up or the backup 120-volt alternating current output now found on many grid-tie inverters as noted in previous paragraphs. Our refrigerator needs about 250 Watts but only runs about 50 percent of the time for an average power need of about 125 Watts over the worse case 16 hour winter night. This needs about two kWh of storage which equates to about two 12-volt marine deep cycle batteries. This battery system should cost under $500 and is a bargain compared to the typical $10,000 home battery backup system or whole house fossil-fueled generator. These two batteries and inverter can provide daily power for weeks with daily solar charging since the lead acid batteries can typically cycle several hundred times before they are worn out.

Although I installed this backup system over five years ago, it has still never been used due to the high reliability of our local grid. I have not even tested it because then I would have to re-set all the clocks in the house due to the momentary loss of power. I am still waiting for an outage to see if the five-year-old batteries are still any good

## The Near Future of Home Batteries

Despite the negative position taken earlier in this book on whole-house batter backup systems, the future may be much different as described in Chapter Six. As this book goes to print, there is much talk about the Tesla Battery GigaFactory that will produce batteries at one-eighth the cost of what the initial batteries for EVs cost only a decade ago in 2008. This drives speculation and the promise of the viability of home battery storage units in the future.

Rising interest in home battery storage is being fueled by media coverage of the great progress being made in batteries as witnessed by the Tesla investment in the Gigafactory. The Tesla "power wall" battery is just the tip of the iceberg with respect to revolutionizing the grid and accommodating renewable energy. As noted in Chapter 6, if the policy, hardware, and entrenched politics of the grid can provide the real-time exchange of electricity at instantaneous value and pricing to match supply and demand to the homeowner, the revolution in batteries will be complete and home energy storage will be both practical and effective.

## Appendix D

## *Choosing Your AC/Heatpump Unit*

Converting an oil or natural gas hot air heating system to a heat pump does not need to be an overly significant investment. The cost of operating it over the life of the system will be about a half to a third of the cost of heating with fossil fuels so whatever it costs, it is still a significant net benefit. For these hot-air HVAC systems, if the duct work is already in place, the only thing that needs to be done is to replace the fossil fuel furnace inside the house with a heat pump indoor heat exchanger and add the outdoor heat pump unit which contains the compressor and condenser coil.

If an air conditioning unit has already been added to older oil- or gas-ducted home systems, the only thing needed is to upgrade the cooling portion of the AC unit from a one-way cooling system to a bi-directional heat pump system that will cool in summer and heat in winter just as effectively. The difference between an AC system and a bi-directional heat pump/cooling system is just a reversing valve that only adds about 20 percent to the cost of the system. In northern U.S. areas, the oil or gas furnace is usually not even removed as it can remain as a backup for very cold nights and maintenance contingencies.

The energy cost with a heat pump system is one-third of that with a fossil fuel system when outdoor temperatures are above 40 degrees and about the same as the fossil fuel system at low temperatures when a heat pump system needs a backup heating system. But the majority of the heating energy in most areas of the U.S. is at less extreme temperatures where they operate most efficiently.

An additional advantage of a heat pump is that it provides air of a comfortable temperature almost continually, where a hot air furnace blows hot air until the air reaches the temperature set in a thermostat, and then the fans turn off, so there is no air circulating until the temperature drops and the fan kicks in again.

Even homes with hot water radiators can switch to heat pumps without needing the typical whole-house conversion to a ducted air-handling system. Contractors give high estimates to convert systems with radiators or baseboard heaters because they assume that you will want both benefits of the heat pump — winter heating and summer whole-house cooling — from the same equipment at no added costs other than the energy to run it. So duct work is usually considered the only viable upgrade path and more than doubles the cost of the investment. But adding duct work might not be necessary, depending on the air-conditioning (AC) needs. If the house has already added window AC units or split coil AC units to meet the cooling needs, there is no need for ductwork. In that case, there are air-to-water or geothermal water-to-water heat pump systems that can replace the fossil fuel boiler with a heat pump indoor heat exchanger and the outdoor heat pump unit. The heat pump then heats the same indoor water for circulating in the radiators and baseboard coils. But radiator heatpump systems cannot be used for air-conditioning and cooling because of the massive amount of condensation that would form throughout the house on the radiators if they were cooled. To get air-conditioning on these systems, window units, mini-split units or an air handler have to be added.

The heat pump only delivers up to about 120°F water efficiently, whereas the old boiler used to deliver 160°F water. This is a significant difference. But even in older homes this might not be a concern if the house has been upgraded to new insulation standards. This usually includes double the insulation in the attic and blown-in insulation in the walls. In this case, the original 160°F fossil-fueled boiler was over capacity and, in many cases, a heat pump circulating 120°F water will work.

## Improving Radiator Efficiency

Fans can be used to improve the efficiency of 120°F water in old radiators that were designed for 160°F water. Our 90-year-old house does not have good insulation because the previous owners' attempt at blown-in insulation was blocked by an internal baffle in the walls that allowed only two inches of new insulation to be blown in through the outside walls. Since we were unwilling to tear up the interior walls to add more insulation from the inside, we chose to improve the efficiency of the heat transfer from the 120°F radiators by using some fans. We put small fans blowing on about half of the radiators that could be hidden behind furniture so they would not have an impact on aesthetics. We wired them to come on when the heat pump is running. On the coldest days we also have a spare box fan or two that can be placed temporarily in front of any radiator to maximize the heat from that radiator in that room.

Another advantage of the fans is keeping the temperature lower in the circulating water of the radiators. If the heat pump has to heat the water to 120°F, it is not operating as efficiently than if it only has to heat to 110°F. By letting the fans run across the radiators, the heat transfer in a room is so much more efficient that the heat pump water rarely gets even to 110 and is thus operating more efficiently. These fans also give us easy zone control over heating. If we are in a room, we turn up the radiator fan in that room and it gets much warmer than other rooms. This way each occupant has direct control over their room temperature by adjusting the fans without having to change the overall house thermostat.

In the past, outdoor air-source heat pumps were originally considered more practical in moderate climates, such as the southern U.S. But higher efficiency models and new refrigerants and newer technology compressors have been coming out for decades that make them practical throughout the U.S.

### Ground-source Heat Pumps

Ground-source heat pumps that use the ground as a source of heat instead of outdoor air are frequently referred to as "geothermal," though the term "geothermal" more accurately refers to scavenging heat from volcanoes and underground hot spots.

Ground-source heat pumps can reduce heating and cooling costs three to four times compared to oil or propane heat because they only have to increase the heat from the constant 45-50°F of the ground, not 20°F in winter for an outside air heat pump. In cooler climates a ground-source heat pump can pay for itself many times over during its 10 to 15 year lifespan, especially when it is time for a new system anyway. And, being electric, it can run from solar, wind, or any other renewable electric source. Never buy another AC unit. Replace it with a bidirectional heatpump/AC unit instead!

### Window AC Units

The disadvantages of window AC units are the loss of window space and leaks from shoddy installation. Most people in older homes have either added central air-conditioning or have a plethora of window AC units for comfort in the summer. While some looked at these AC units as a stop-gap measure, these days one can take a different view and consider their advantages. Window or room AC units are the ultimate in modern zone controlled cooling. You simply turn them on in a room that needs cooling or turn them off when not needed.

If you have taken the advice to replace these air-conditioning-only units with combined heat pump AC units, then you also have the ultimate in zone controlled heating. Just turn it on in the winter as a heat pump to boost clean, renewable, fossil-fuel-free heat in any room where it is needed and at one-half to one-third the energy use.

### Portable AC Units

There are a plethora of "portable" AC units now on the market, but most of the ones I have seen are used in the most inefficient manner. There is very little that makes them "portable" because they need one or two large hoses to connect to the outside air for heat exchange. And these hoses need to be installed in windows tightly to prevent leaks. To operate properly, there needs to be good thermal isolation between the indoor condenser coil and the outdoor evaporator coil for maximum transfer of heat from inside to outside.

Single hose portable units are especially inefficient because they use cooler indoor air to blow across the hot evaporator coils and then exhaust that air outdoors. The air to replace that cool indoor air has to come from somewhere and it comes from the outside through all the leaks in the rest of the house. The portable unit provides cool air where it is blowing, but it is at the expense of bringing hot, humid outside air that warms the rest of the house due to air incursion.

Dual hose portable units are much better, but more cumbersome to install. They use only outside air coming in one hose and out the other. Thus the cool air stays inside and no outside air is pulled into the rest of the house so it operates rather efficiently. There is one further disadvantage of the portable unit. Because the compressor is also in the same box indoors, the noise is inside so be sure to get a quiet one. Heat generated from running the compressor indoors cancels some of the desired cooling effect when in A/C mode. The product is good, but improper installation can undermine much of the efficiency gained.

For a little more investment, these portable A/C units can be purchased with a reversible heat pump function as well. If installed properly, reversible combined A/C-heat pump portable units with dual hoses have advantages. They are heat pump systems that can produce either heating or cooling on clean renewable electricity. The disadvantage of the A/C compressor being indoors is an advantage during the heating cycle, because that heat stays indoors where it adds to the heat delivered by the unit.

### Combined A/C-Heatpump Units
Even if you heat with fossil fuel, never buy another central A/C unit without at least considering buying a one-to-one replacement with a combined A/C/heat pump. It is

Portable

Window

Combined Air conditioning
and Heatpump Units

Mini-Split

Figure 36. Heatpump options for both heating and cooling in one appliance.

far better to replace the A/C only unit (including window units) with heat pump units that can not only continue the needed cooling, but provide heat in the winter at half the cost of oil, gas or resistance electric (*Fig. 36*). There are simple low cost ($1500) mini-split units that don't even need a window. The small pump unit goes outside and the inside blower can be in any room of the house. And it needs no duct work. The indoor coils of these units can be placed almost anywhere and are then connected by only a thin convenient run of two copper pipes to the companion outside unit which houses the compressor and outdoor coil. This too can be placed in an inconspicuous location to retain the aesthetics of the old house. There are even models that fit in the ceiling. They look not unlike a simple ceiling duct that is flush with the ceiling since the coils are recessed into the attic between the joists. They require no ductwork because they are connected to the outdoor unit by just the copper pipes. One outdoor unit can serve two or more indoor units to make sure the conditioned air, either heating or cooling gets to the right locations.

Never buy another A/C unit again. When it is time to repair or replace a window A/C unit, a whole-house A/C unit, a portable unit, or a mini-split A/C unit, the added cost of choosing a bi-directional heat pump that can provide both heating and cooling from the same unit is only a small percentage. Paying just a little more to include the reversible heat pump function can save thousands in your heating bills over the life of the unit, not to mention reducing your carbon footprint drastically by not burning any fossil fuel during most non-extreme winter days.

# *Endnotes*

Websites accessed March 1, 2018.

1) Guardian, 2015. <theguardian.com/environment/2015/sep/29/carney-warns-of-risks-from-climate-change-tragedy-of-the-horizon>.

2) Mark Frauenfelder, 2008. <boingboing.net/2008/03/11/all-the-water-and-ai.html>.

3) U.S. Geological Survey. *The World's Water.* <water.usgs.gov/edu/earthwherewater.html>.

4) Hooke, Roger LeB., Martin-Duque, Jose F., 2012. Land Transformed by Humans: A Review *Geological Society of America* <geosociety.org/gsatoday/archive/22/12/article/i1052-5173-22-12-4.htm>.

5) World Bank, 2014 <data.worldbank.org/indicator/EG.USE.ELEC.KH.PC?locations=US>

6) Clean Technica, 2014. <cleantechnica.com/2014/09/04/solar-panel-cost-trends-10-charts>

7) "Prepared mind" Calvin W. Schwabe, Pendle Hill Pamphlet #343 *Quakerism and Science*, p26.

8) Dreby, Ed, and Keith Helmuth, 2009. *Fueling Our Future.* <quaker-books.org/fueling_our_future.php>.

9) U.S. Energy Information Administration. *Today in Energy* <eia.gov/todayinenergy/detail.php?id=30112>.

10) U.S. Energy Information Administration *Short-term Energy Outlook* <eia.gov/forecasts/steo/report/coal.cfm>.

11) As Electric Vehicles Gain Favor, Utilities Can Accelerate EV Adoption, February 15, 2018 by American Council for an Energy-Efficient Economy (ACEEE) <theenergycollective.com/aceee/2423645/electric-vehicles-gain-favor-utilities-can-accelerate-ev-adoption>.

12) World Bank, 2014 <data.worldbank.org/indicator/EG.USE.ELEC.KH.PC?locations=US>.

13) The discussion of "peak oil" has now shifted from the original concept of the oil supply running out which would drive costs ever upward. The new economic concern is over declining demand for oil which results in the loss of value to the fossil fuel industry, which is a cause of great concern to economists.

14) O'Connor, Peter, 2017. Solar Jobs, Coal Jobs and the Value of Jobs in General. *Union of Concerned Scientists* <blog.ucsusa.org/peter-oconnor/solar-jobs-vs-coal-jobs?_ga=2.162568060.1759622519.1519912895-10501845.1519912895>.

15) How Many Trees? <sciencefocus.com/qa/how-many-trees-are-needed-provide-enough-oxygen-one-person>.

16) Pierre-Louis, Kendra, 2017. *This is What It Looked Like Before EPA Cleaned It Up* <popsci.com/america-before-epa-photos#page-5>.

17) Technological change incited by EPA regulations has yielded reductions of 70-95 percent of air pollutants relative to 1970s emission

rates. Just since 1990, carbon monoxide is down 77 percent; lead is down 99 percent, nitrogen dioxide is down 47 percent, ozone is down 22 percent; and particulate matter is down 37 percent—all below the current standards for each pollutant <gispub.epa.gov/air/trendsreport/2016>. These dramatic improvements in air quality have happened despite substantial increases in population, number of vehicle miles driven per year, and GDP. Many people are not aware of this air pollution control success brought about by EPA regulations, the same regulations that are now being eliminated by the Trump Administration as impediments to the future of the fossil fuel industry <whitehouse.gov/the-press-office/2017/01/30/presidential-executive-order-reducing-regulation-and-controlling>. To head the EPA Trump appointed Scott Pruitt, who has filed 14 suits against the EPA when he was Oklahoma Attorney General <businessinsider.com/trump-epa-pick-scott-pruitt-2017-1>. Pruit denies that human activity and CO2 emissions are responsible for climate change. He is determined to roll back EPA regulations that protect our clean air and water. The 2018 budget proposed by the Trump administration slashes the EPA's ability to enforce clean air and water regulations by 60 percent <edf.org/blog/2017/04/12/latest-polls-trumps-environmental-agenda-collision-course-voters>.

18) Environmental Protection Agency, *Environmental Studies: Documerica by Location* <archives.gov/research/environment/documerica-geographic.html>.

19) Lawrence Livermore Laboratory, *2016 Energy Flow Charts* <flowcharts.llnl.gov>.

20) National Academies of Science, Engineering and Medicine, *Energy Efficiency in Transportation* <nap.edu/read/12621/chapter/5>.

21) National Academies of Science, Engineering and Medicine, *Energy Usage in the U.S. Residential Sector, 2015.* <needtoknow.nas.edu/energy/energy-efficiency/heating-cooling>.

22) National Academies of Science, Engineering and Medicine, *Energy Efficiency: Heating and Cooling* <needtoknow.nas.edu/energy/energy-efficiency/heating-cooling>.

23) U.S. Energy Information Administration. *Today in Energy, February 27, 2017.* <eia.gov/todayinenergy/detail.php?id=30112>

24) Roberts, David, 2017. The Key to Tackling Climate Change: Electrify Everything <vox.com/2016/9/19/12938086/electrify-everything>.

25) Energy Collective, February 15, 2018 <electricauto.org/?page=EVHistory> <mashable.com/2017/10/03/electric-car-development-plans-ford-gm/#qn2vmMPdqiqy>.

26) Department of Energy, 2014. *The History of the Electric Car* <energy.gov/articles/history-electric-car>.

27) EV Misinformation Abounds! A Battery is not a Tank <aprs.org/EV-misinformation.html>.

28) Solar Journey USA <solarjourneyusa.com/EVdistanceAnalysis.php>.

29) Ford Study <greencarreports.com/news/1099531_electric-car-drivers-tell-ford-well-never-go-back-to-gasoline>.

30) Make Wealth History <makewealthhistory.org/2017/04/24/what-happened-to-swappable-batteries-for-electric-vehicles/>.

31) World Economic Forum *Accerlating Climate Action* <weforum.org/agenda/2017/09/countries-are-announcing-plans-to-phase-out-petrol-and-diesel-cars-is-yours-on-the-list>.

32) NY City auto disruption <mountaintownnews.net/2015/08/20/tony-sebas-startling-view-of-market-disruptions>.

33) No Horses <mountaintownnews.net/wp-content/uploads/2015/08/5th-ave.jpg>.

34) Casey, Tina, 2017. *Under Trump's Watch, Wind Energy to Blow Past Coal in Texas* <cleantechnica.com/2017/12/04/trumps-watch-wind-energy-set-blow-past-coal-texas>.

35) Durrenberger, Mark, 2015. Solar vs Trees <newenglandcleanenergy.com/energymiser/2015/09/24/tree-math-2-solar-vs-trees-whats-the-carbon-trade-off>.

---

36) *Community Solar Resources and Examples:*

Solar Energy Industries Association (SEIA) <seia.org/policy/distributed-solar/shared-renewablescommunity-solar>.

Solar Oregon community solar project <solaroregon.org/how-to-go-solar/solar-for-communities>.

Oregon Clean Power Cooperative, http://oregoncleanpower.coop/Solar.

Washington community solar project <solarwa.org/community_solar>.

Community Solar California <communitysolarca.org>.

National status of community solar Reference – Solar Energy.

Industries Association (SEIA) <seia.org/policy/distributed-solar/shared-renewablescommunity-solar>.

There are currently 26 states with at least one community solar projects on-line. The combined power rating of these 101 projects is 108.48 MW.

The states having the largest number of on-line solar projectsinclude Colorado (43), Massachusetts (11), Minnesota (7), and Washington (6).

Community Solar California <communitysolarca.org>.

Community Solar - Oregon , Solar Oregon community solar project <solaroregon.org/how-to-go-solar/solar-for-communities>.

Oregon Clean Power Cooperative <oregoncleanpower.coop>.

Corvallis High School Solar Project, Corvallis OR --- 117 kW system –Completion slated for summer 2017.

First Unitarian Church Portland Oregon --- 38 kW system – Completion slated for summer 2017.

Mazama Club House, Portland OR <oregoncleanpower.coop/wp-content/uploads/2017/07/20170706-mazamas-solar-pr.pdf>.

Rouge Valley Council of Governments Solar Project—37 kW system –Completed April 2017.

Solar Washington community solar project <solarwa.org/community_solar>.

Avista, Spokane WA 425 kW Project In process, completion date unspecified <avistacorp.mwnewsroom.com/News/in/Avista-Announces-Community-Solar-Program>.

Clark County WA Public Utilities — 319 kW project, completed August2015 <clarkpublicutilities.com/community-environment/what-we-do/green-programs/community-solar>.

Skagit Community Solar, Anacortes WA <skagitsolar.org/projects.html>.

Five systems in Anacortes and La Conner totaling 85 kW completed-between 2013 and 2015.

Woodstone Community Solar Project, Bellingham WA — 46 kW system– Completed Summer 2014 <woodstone-corp.com/wood-stone-solar>.

37) Pamela Haines, Ed Dreby, David Kane, and Charles Blanchard, 2016. *Toward a Right Relationship with Finance.* QIF Focus Book #9. Quaker Institute for the Future <quakerinstitute.org>.

38) The Telegraph, 2014. Solar Panels Better than a Pension. <telegraph.co.uk/finance/personalfinance/pensions/10615852/Solar-panels-better-than-a-pension-says-minister.html>.Heating oil with 139,000 BTU/gal at 70% efficiency is about the same 100,000 BTU as 30 kWh at ten cents or $3.

39) Quaker Earthcare Witness <quakerearthcare.org/search/site/fracking>.

40) Fox, Justin, 2017. The Energy-Changing Power of the LED Bulb <bloomberg.com/view/articles/2017-05-09/the-economy-changing-power-of-the-led-bulb>.

41) EPA Statistics: Gas Mowers Represent 5% of U.S. Air Polution <peoplepoweredmachines.com/faq-environment.htm#pollutants>.

42) Banks, Jamie. *National Emissions from Lawn and Garden Equipment* <epa.gov/sites/production/files/2015-09/documents/banks.pdf>

43) *Ibid.*

44) Ben Jervey, *Want to Improve Wind and Solar? Bring Them Together.* <ensia.com/articles/renewable-energy-wind-solar>.

45) U.S. Energy Information Administration, April 11, 2017. California's Rising Solar Generation Coincides with Negative Wholesale

Electricity Prices <cleantechnica.com/2017/04/11/californias-rising-solar-generation-coincides-negative-wholesale-electricity-prices>; <eiz.gov/electricity/data/eia861m/index.html>; <caiso.com/market/Pages/ReportBulletins/DailyRenewablesWatch.aspz>.

46) Denholm, Paul, Matthew O-Connell, Gregory Brinkman, and Jennie Jorgenson, 2015. Overgeneration from Solar Energy in California: A Field Guide to the Duck Chart <nrel.gov/docs/fy16osti/65023.pdf>.

47) Tesla Switches on the World's Largest Lithium Ion Battery <mashable.com/2017/11/30/tesla-battery-powerpack-south-australia/#d6gCvufRIOqk>.

48) Bruninga, Bob. *My first EV, The Elect-Reck at Ga-Tech*—1970 <aprs.org/EV-at-tech.html>.

49) My Solar Experience: <aprs.org/solar.html>.

50) Annapolis Friends Meeting Environmental page: <aprs.org/AFM/environment.html>.

51) Annapolis Friends Meeting EV charging signs.<aprs.org/EV-charging-signs.html>.

52) DOE Workplace Charging Challenge <energy.gov/eere/vehicles/ev-everywhere-workplace-charging-challenge>.

53) Anapolis Friends Meeting Volunteer Interfaith BikE Sharing. VIBES <aprs.org/vibes.html>.

54) Swennerfelt, Ruah, 2016. *Rising to the Challenge: The Transition Movement and People of Faith*. QIF Focus Book #10. Quaker Institute for the Future <quakerinstitute.org>.

55) Tesla Pickups <frogcars.com/2018-tesla-pickup-truck>.

56) BMW EVs <electrek.co/2017/09/07/bmw-updates-ev-plan-electric-cars-mass-production>.

57) Mercedes <engadget.com/2017/09/11/mercedes-benz-electric-versions-2022>.

58) Ford <freep.com/story/money/cars/ford/2017/01/03/here-details-fords-electric-vehicle-plan/96109758>

59) Volkswagen <usatoday.com/story/money/cars/2017/09/11/volkswagen-mercedes-benz-electric-cars/653903001/>.

60) General Motors <detroitnews.com/story/business/autos/general-motors/2017/04/21/gm-electric-china/100761754>.

61) Honda <businessinsider.com/honda-to-launch-2-electric-cars-by-2018-2017-8>.

62) Toyota <nextbigfuture.com/2017/08/toyota-plans-to-leapfrog-tesla-electric-cars-by-2022-with-fast-charging-solid-state-batteries.html>.

63) Nissan <greencarreports.com/news/1112706_nissan-mitsubishi-renault-to-launch-12-new-electric-cars-by-2022>

64) Hyundai <greencarreports.com/news/1112165_hyundai-kia-eight-electric-cars-by-2022-dedicated-ev-platform>.

65) Volvo <fortune.com/2017/07/05/volvo-electric-cars-hybrid-2019>.

66) Electrical Vehicle Association of Greater Washington, D.C. EV Information Sheet <evadc.org/wp-content/uploads/2016/01/EVInfoSheet-20160118.pdf>; Compare EVs <insideevs.com/s-crosses-10-billion-e-miles-driven-facts-and-graphs>; <usatoday.com/story/money/cars/2017/09/11/volkswagen-mercedes-benz-electric-cars/653903001/>.

## About the Author: Bob Bruninga

I hope I have been able to provide enough background on this fascinating birth of clean renewable energy in our time to point seekers toward the right answers in their search for integrity in our use of energy and stewardship of our resources. I am an engineer and can only speak to the practicalities of energy details. I will leave the elocution of the spirit and mindful benefits to other Friends who have well covered these topics before me.

My career in electronics, communications, and space began in 1955 with the building of little telegraphs from images in the encyclopedia in second grade. By fourth grade all the kids within three blocks were networked. And by middle school (1957) the space race was on. By high school we had ham licenses and could communicate great distances using WWII surplus radios. To power these toys during scout camping I rewound an old car generator to produce 110 volts from a discarded lawnmower motor.

This early interest in energy evolved to my senior project at Georgia Tech in 1970 to electrify an old VW to participate in the first MIT/CalTech Cross-country Clean Air Car Race. Though the car only made it to the Mississippi, I was shipped across country to Navy graduate school in Monterey, California. During these years, Amateur Radio was exploring the use of VHF radios for local two-way communications. At my first duty station in Pearl Harbor, I configured a suitcase-sized radio in the trunk of my car and used an old telephone dial to make phone calls through the Ham radio repeater on Diamond Head in 1971.

During this time the University of Hawaii had experimented with similar radios to interconnect remote teletypes from the other islands to access the central computer on Oahu. This shared use of a data channel was called the "Aloha protocol" and can be considered the beginnings of what is now the Internet. While stationed in the Washington D.C. area, by 1975 our local radio club had developed similar teletype links to one of the first microcomputers (a 6800) in my basement. We spun off teletype support into one of the first dial-up TTY networks for the deaf. For faster speed, the local club (AMRAD) developed the AX.25 protocol for data over radio. The FCC made it legal in 1978 and it is viable to this day.

During a sea tour in Japan I began to experiment with this data channel for plotting the location of ships and by the mid-1980s had developed a protocol to facilitate Ham radio support of emergency communications. By 1992, I combined the position information with the traffic data and called it the Automatic Packet Reporting System (APRS) to match my call sign, WB4APR. Since that time, APRS has grown worldwide with many tens of thousands of users.

At the Naval Academy in 2001 I married a lovely astrophysicist and got access to her observatory roof for my antennas. I mentored my first student satellite to carry this protocol into space. This was followed by ten other such spacecraft and experiments.

In 2007 I had my energy epiphany and began my remaining lifelong leading to change the world to renewable energy as described in these chapters. Since my day job for the last two decades has been in aerospace and space systems, it was a real eye opener to see how critical every aspect of spacecraft design is to the overall performance of the systems on board when spacecraft are in the void of space. The surface color of a sphere in space can change the steady state temperature from a very cold – 60°F for a white sphere, to a cool 55°F for black, to over 250°F for a shiny aluminum surface. This wide temperature range shows how critical is the exact condition of the planet in space at this instant in geological time to the amazing balance that sustains life on Earth. It is our atmosphere and surface characteristics that defines the "color" of Earth with respect to solar heat gain and thermal heat loss, which then defines our steady state temperature.

Our atmosphere and life on this planet are totally intertwined. For most of the 4.6 billion years since the formation of Earth, the atmosphere was nothing like it is today. It began with a toxic brew from all the volcanic activity that could not possibly support life as we know it. But eventually life itself began to assemble molecules and gather energy to sustain itself. These original living cells soon included cyanobacteria, which produced oxygen from carbon dioxide and water through photosynthesis. Eventually other forms of life evolved that live on oxygen, and the balance of gases in the atmosphere began to support life as we know it today.

The era of humans is only a blink of the eye in evolutionary time. As we poison our atmosphere and change its balance, so also will the temperature, weather, plants, and animals change. In the past, these changes were evolutionary, over tens of millions of years. Now, however, we are upsetting the balance of carbon, which took millions of years to sequester out of our atmosphere, by burning it back into our atmosphere in our insatiable quest for more energy. We are seeing rapid changes in the atmosphere that our own descendants in only a generation or two will see and remember that we were responsible.

# Quaker Institute for the Future

*Advancing a global future of inclusion, social justice, and ecological integrity through participatory research and discernment.*

The Quaker Institute for the Future (QIF) seeks to generate systematic insight, knowledge, and wisdom that can inform public policy and enable us to treat all humans, all communities of life, and the whole Earth as manifestations of the Divine. QIF creates the opportunity for Quaker scholars and practitioners to apply the social and ecological intelligence of their disciplines within the context of Friends' testimonies and the Quaker traditions of truth seeking and public service.

*The focus of the Institute's concerns include:*

• Moving from economic policies and practices that undermine Earth's capacity to support life to an ecologically based economy that works for the security, vitality and resilience of human communities and the well-being of the entire commonwealth of planetary life.

• Bringing the governance of the common good into the regulation of technologies that holds us responsible for the future well-being of humanity and the Earth.

• Reducing structural violence arising from economic privilege, social exclusion, and environmental degradation through the expansion of equitable sharing, inclusion, justice, and ecosystem restoration.

• Reversing the growing segregation of people into enclaves of privilege and deprivation through public policies and public trust institutions that facilitate equity of access to the means life.

• Engaging the complexity of global interdependence and its demands on governance systems, institutional accountability, and citizen's responsibilities.

• Moving from societal norms of aggressive individualism, winner-take-all competition, and economic aggrandizement to the practices of cooperation, collaboration, commonwealth sharing, and an economy keyed to strengthening the common good.

CPSIA information can be obtained
at www.ICGtesting.com
Printed in the USA
BVHW061833110220
572026BV00021B/2027